INTRODUCTION

AF884366

It may sound richly over-stated to say that our own reality must be perceived through one of two contrary perspectives; yet this is the case. There are many who hold to and live by a temporary perspective of life and reality. And the fewer who hold and live through an eternal perspective of the same. Which do we each hold, and why? Which should we hold to, and more importantly, why? What are the core reasons and purpose of life we each see and hold to, and why do most all souls see life and all things in so many individual temporary ways? All these questions share a few common answers; as a true quest of eternal importance has been laid out before us all. It is all about where we are eternally going, what we currently are, from where we all have come, and what we should all be. We had a true start to *our* awareness; but that cannot in any way mean we can ever have a conscious end. We have many interruptions and voids in our awareness to be sure; and our core desires which force our unfolding choice of direction. But a true eternal end to our awareness can never come. To explore, define, and explain, all of these existential qualities, we must discuss the many ways, means, and aspects of all correct, and contrary desires. And all in our hope of igniting the bright eternal perspective in you.

Author, Lawrence Michael Stack: Theorist, Philosopher, Metaphysician, and Theist.

Friends call me Michael.

Copyrights© July 18, 2021 Lawrence Stack Publisher

All rights reserved. No part of this book may be reproduced, stored in retrieval system, or transmitted in any form or by any means, electronic, photocopying, recording, or otherwise without licensed consent; with the sole exclusion of all public domain KJV Scripture quotes, which are identified in quotation marks, and being in all italicized capitalized fonts.

ISBN 9780578350844

LCCN 2021925909

Other *currently* available books by this author…

1. The Theory of Created Conscious Perception.
2. All Conscious Formulations.
3. Live to know the GOD of Prophecy.

Lower case italics, implies mild added emphasis.

To see *and* be seen by GOD

This book is in continuation of "The Theory of Created Conscious Perception" which we hope you have read first, to adequately hear all we express. All souls, and life itself, are set in projecting our own force out from our own conscious and physical center of reality. We know this firmer realm's four forces, originate from the one supreme conscious source of desire and intent, as do the ruling forces of the higher vast ethereal realm of Heaven. We see all laws, principles, and theories, and reality itself, and even force at some reductive point, must be quantized, basic, sourced, and real. There can be no endlessly reducible reality, or reality can never be defined, and does not exist. We state that by our two strict uses of the term "Perspective" we show and define that there are only two available forms of core belief we can hold concerning our place and purpose in reality. These two primary perspectives, or directions of understanding and belief, are the eternal perspective of life, and all temporary perspectives of life. Our perspective controls how we see and approach every aspect of life, including GOD. An eternal perspective sees GOD as vastly superior to one's self in every way, and to be lovingly obeyed without question. Temporary perspectives see GOD in lesser ways, and more, as reflecting their own desires. The eternal perspective is built in a pure spiritual way as our anchoring foundation. While temporary perspectives offer no understanding to break free from seeing the physical side, as the primary side of life. The eternal perspective sees life as a spiritual platform, with its lesser physical aspects. Temporary perspectives see life as a being a physical platform with their added lesser spiritual aspect. A true perspective is bound to GOD. Temporary perspectives are bound to the flesh and the earth. We ask you to accept, that to properly discuss the degrees and strengths of these two perspectives, and what they essentially are and mean. We must see and understand the underlying ways and means in which these chosen perspectives are formed in all souls. We must know all spiritual *and* physical foundational truths, and how they affect our choices in forming our agency of self-awareness. Now at the very beginning of all complex reality, we see how force rules in creating all things, and *nothing* physical or complex are formed, energized, repelled, drawn, or moved, without a first affect and effect of force acting upon it, or within it. And by this immutable fact, we know our created perceptions are *forcefully* formed as new created realities. So, all perspectives, laws, and forces of consistency in each realm, must be created *and* perceived, relative to the graduating size and speed of its gifted souls, and their small firm, or vast ethereal bodies. Now is the quantum classical divide singular, or is it part of a spectrum of realities? Have any quantum aspects of the higher ethereal realm of Heaven ever been detected? Is there a higher faster, yet undetected electromagnetic spectrum in the overlayed ethereal realm of Heaven. Materialist can't detect or reconcile the higher faster forces that cause complex quantum entanglement at a great distance. By projected conscious force creation, *all* particles that exist are forcefully entangled at some level as everything falls under desired purpose. Please see the created electrons and lesser point particle aspects forming an electromagnetic field around a bar magnet, are consciously set and aligned in a collective effect of one forcefully *entangled* purpose of direction at both ends. Force is synonymous with projected purpose, direction, and guidance. *Our* believed perception*s* of creation, and creation*s* of perceived belief, are experienced within this slow small firmer realm of laws. Are there realms of perceivable eternal beliefs below ours? No. And by seeing only this slower small realm's inflating remnants and ruins, materialist can believe reality started from a pure physical miracle, being here of itself, minus GOD. This "Principle of

Desirous Perception and Belief" has always been, or it could not exist now in our souls. So, this external layered creation of FATHER, can exist only until it loses its purpose. Who can desire and choose to believe that all force and the physical, just emerged from nothing and nowhere; when all common and higher senses, say both have always existed. We see ongoing created and destroyed spherical complexity; but we find not one linear aspect in reality. An unintentional miraculous physical beginning has no basis in science, or natural law, to support it; and not one fine thread of empirical proof. Mankind has not one equation that disproves GOD as the source of reality. And minus their physical miracles, they can't begin to explain the existence of anything. All physical reality declares that for these layers of complexity to exist now, their guiding laws and means, had to exist prior to their current created configuration. So, now there are two systems of faith and belief: one being, first there was nothing, or one singularity, that then beyond all explanation, exploded into endless physical universes, filled by an enigmatic appearance of conscious awareness. Or two, there has always been the desirous conscious forceful projected intent, over the purely physical aspect. Materialist see an unintended reality, born of physical miracles; fashioned from a *temporary* perspective opposed to any Godly purpose or eternal perspective. Now both perspectives see all physical creations now exist in devolving states; but what could they be devolving from? We all know things can only *devolve* from a superior state. So, reversing this trend, back to a beginning, we find something once perfect. Souls with a strong godless desire, can conceive of a false reality, using extravagant mathematical formulas, tricks of light, asymmetrical beliefs, and all the assumptions that arise after accepting lesser probabilities and assumed physical miracles. How can such an afront to science, be the foundation for all their secular delusions? They hide many unscientific skeletons in the back of their mind's closet. It's a poison pill they have all swallowed. It's a non-science that can never be empirically evidenced. How can a great belief in purely physical miracles, be the foundation for all that they theorize? Such a twisted paradox, such a blind contrivance, seeing their greatest commandment, is that belief in GOD or miracles, is a false comfort for ignorant superstitious people. How does science define a miracle? They say a physical miracle is anything that cannot be explained by known laws, and then physically manifests contrary to the laws of physical cause and effect. So, we challenge any scientist to in person and in public, explain and show us all a first causal source for your theories, without using a physical-miracle. Deny all the miracles in your Big Bang, Abiogenesis, Evolution, and physical awareness. Their theories are filled to the brim with physical miracles. If I were of financial means, I would offer an immense reward, for any inarguable proof of any non-miraculous, purely physical start of reality. Their accepted miracles are godless gods, that comfort their own ignorant superstitions. Why do they need and accept so many miracles in their physical beliefs. Yet they deny the forceful patterned projections of a conscious Creator, not wanting to be subject or accountable to a higher conscious source of existence. They accept any alternative, no matter how impossible it may be. They blindly accept endless miracles without blinking, and build unstable beliefs upon them. From the very first to the very last of everything, there must be a pure consonance of one founding truth that runs through all the conscious and physical aspects of all realities. I say no matter who the pure truth offends, no matter how much it offends them including myself, the complete honest truth is everyone's eternal right to hear and accept, or reject. And if rejected, one temporarily believes and lives as one desires. But none has the right to silence or forestall the truth from being heard by all. By the gift of GOD to you of your free accountable soul, you can choose to live your life in all the contrary ways you please; and you will suffer in those devolving ways forever. But if you in any way deny, or inhibit others from hearing GOD'S truth, you must suffer a manifold judgment of never-ending regrets. All must weigh this cost, because all accounts will be called, and all debts will be paid forever. Do not doubt that your free-will came as a gifted sacrifice out from FATHER; yet by the laws of love, it is a gift that has its never-ending accountability. Do not say you did not ask to exist, that matters not. And now evil forces come in every projected divisive desire to exist in ways contrary to GOD. Again, can any publicly, not just disagree, but prove me wrong. I need not hear about other consensus beliefs, or hear of your deep hopes and desires. I've heard them all before, many times. Do not meander about in a long silique of tangled words and deft reiterations. Do not fill a room with confused souls who applaud *your* beliefs. I say show and prove to *me* what force actually is, and from what source does force originate. Show me from where the physical building blocks of matter originate. Show me your applied reasons for any certain quality and form to exist that supports your belief, without the endless problems your belief is plagued with. We see that the prior existence of force and matter just existing in an impossibly condensed singularity in the middle of nothing, is in all ways contrary to science, beyond all odds, and requires a firm definition of nothing. Can I scientifically define what is nothing, or miraculous, for everyone. Well first, in that by the existing conscious creative power of GOD, there is no such thing as the miraculous or unexplainable. All things imagined by FATHER, represent his thoughtful considered desired intention. And at his considered strong knowing desire of looking and speaking forth into his aether, his forceful patterned desires project forth, to create FATHER'S external three-dimensional perceived certainty. Reality being created by FATHER'S desire is thereby explained. Miracle, means an inexplicable manifestation; and physical creations are only unexplainable minus conscious force projection into the aether. Without GOD being eternal before us, our own perceptions are unexplained manifestations. And as for what is nothing; I see that all that GOD and we currently perceive, currently exists; and nothing means all former things that no longer exist, and all coming things that do not yet exist. All that is currently beyond the projected desires and intent of GOD and us, is nothing. We see the desires of GOD create all true realities in consonance. And only godless scientific delusions exist as unexplainable manifestations, or miracles.

Now the term GOD, can also mean not just one soul, such as FATHER or JESUS. It can mean the one congealed mind of belief and love forged between the Trinity and Holy Angels and all the Holy Saints. If we are one with GOD, the Bible says we are lesser Gods. Without doubt, our every projected desire has its projected purpose and intention. Can many accept these true definitions, or only a few? And here I bring up the greatest of all desired forceful purposes, born of the greatest needful intentions, being love. Love is the force that wants, needs, and commonly draws, orders, aligns, builds, raises, secures, nourishes, provides, and sustains. And those souls, consciously physically existing in free-will rebellion to GOD'S love, are a rising forceful antithesis to every loving binding desire of GOD. Now seeing FATHER has only omniscient omnipotent omnipresent desires and intentions, should he change his desired intentions to align with the differing beliefs in each of the billions of individuals? That is an untrue question, yet it has a most true answer, being NO! It is we who are to learn to commonly understand and align our desires and intents to those of our loving providing FATHER. We all know hate is the antithesis of love, and that where love builds, hate tears down, love adjoins and hate divides, love creates and hate destroys. Hate is individually defined as any desire contrary to one's love; and so, it is felt as a force against, one's love. So, between all individuals, every contrary love is felt as hate against all differing loves. If mine and your love are contrary, my love is felt by you as hateful, and your love is felt as hateful towards mine. So, individual contrary understandings and beliefs of love, can never live in peace. Our knowing and growing one common love under the omniscience and omnipotence of GOD, is the only way that love between many can work for eternity. So, we see our understanding of desire and intent, is required to hold an eternal perspective of reality that focuses on and seeks to sense all things through GOD'S common spiritual bond without end. In quickly glimpsing the sum of creation, we know it was originally created to bring to GOD a great family; and to eternally sustain that family. But we see this creation being destroyed, and the hate of contrary loves is the cause. All loves, not aligned to GOD, are hateful towards GOD. If things do not align for, they are against. Hate is any projected force of desired intent, not aligning with GOD'S created ways. All evil is born of and resides in what is called individualism; or individual beliefs. There is a battle between all individual loves, occurring within a battle of love for and from GOD. For literary house-keeping, we remind all that antithetical forces can only exist among more than one distinct conscious soul; one soul cannot hold two contrary *equal* core loves. Now every individual's highest love for anything other than GOD must be temporary, as it must exists in ever-changing unsustainable ways. We clearly see how the rising number of souls, adds to the divisions and degrees of insanity. And in *our* struggle, as FATHER'S Children, we see his Holy Laws and Rules apply to us. A conscious battle field now exists, and so the questions arise; who created the field on which this battle started and now rages, and who made the rules? Who started this war? And if one supreme soul created the field, the rules, and the participants; are they not ensured a victory? Yes. How could FATHER, having omniscient understanding, and the force to perceive this vast complex search for love, create a process that could overthrow his purposeful intent. Who now projects all these antithetical forces into this battlefield? It is all who do not emulate CHRIST. And towards FATHER'S purpose, to what extent can many free-willed souls joyfully become *one* in his pure belief and purpose. We see among CHRIST, HOLYSPIRIT, and all the Holy Angels and Saints, that they all desire to hold, share, and build upon FATHER'S desired understanding and belief with others. But among Satan, his fallen angels, and fallen mankind, we see desires to build many contrary understandings and belief. Yet, in all this, no soul was ever diminished in their self-awareness or free-will agency of choice. Souls can never lose their agency to any belief, they can only desire and adopt a belief. It is just a matter of each soul choosing GOD'S love and desire, or a contrary belief. I have heard souls say following CHRIST, would mean giving up all that they are, to then be nothing. No, it means changing all you are through love, choice, and new understanding; and fighting through the elements of chance occurrence along the way. We see endless *individual* desires exist and are expressed in each and every way that fallen souls choose to live and believe. And in this, their temporary perspective is born, being a perspective focused on current non-spiritual physical moments, places, and experiences, that continually change by choices and chance. Do you see the level of created complexity, required for FATHER to create choice and chance? Do you see that even our complex DNA, and our formed desires in propagation, are all part of manifesting our true choices and all chance occurrence? We must see how our greatest core desires get stronger or weaker with our ongoing choosing and believing process. Now contrary love, is a force people project and sense, often not seeing it as hate. Hate is every desire and belief that is contrary another's desires and beliefs. And there are of course degrees of hate. And most find the endless minute levels of differing desires between them and others to be tolerable in this world of individuals. These minor levels of hate are not considered as hate, or hateful by most. It is only after levels surpass the point of one's personal tolerance level that a disagreement is sensed as being hateful towards them. Thus, we see the insanely sensitive social justice warrior arise, as a generation of highly-sensitized souls have been indoctrinated. Each soul has their own parameters, as to what they see as hate from, or towards them. So, we see some as too sensitive, or not sensitive enough. Love, and non-love or hate, in their degrees, are projected forces of desire and intent to exist in certain ways. The point is, that feeling hate from another, is sensing their strong contrary force of belief and love, *against* your soul. Contrary beliefs are far more divisive than most believe. An easily *tolerated* measure of contrary-love is seen among most as no more than a common difference of opinion between souls concerning the ways of living life. But when we add GOD'S desires and ways of *eternal-life* into this formulation, for which this creation was designed. It is then GOD whom we all lovingly agree with, or hatefully disagree with in our own degrees. Souls sense hate as a forceful desire from another, that is beyond their level of tolerance. So, these play a god in defining good and evil, and love and hate. Now all false temporary perspectives of life, desire and need to sense they are broadly accepted. But let me momentarily pull you over into the perspective of eternal life and peace. Do you see a difference in the equation for an eternal love and life, and a temporary love and life. Do you have hope in an afterlife of pure loving peace in a perfect physical estate called Heaven. We must see how the tolerable level of hate within individuals varies greatly. And it is all the individual ways of life that brings all hate in this world. Some souls insanely riot and kill, as others wonder why they riot and kill. All temporary individual perspectives, breed aggression, violence, hate, and death, and they underly all physiological and psychological issues. Can such contrary individual approaches of belief towards life, ever be brought into the last eternal stage of loving peaceful joyous reality? Can this world's failed doomed individual approach to life be brought into Heaven, minus a horrid result? FATHER must replace the current Earth *and Heaven*, as it too is defiled by the sin of fallen angels. So absolutely not. Our inherent free-will agency of awareness is *eternal* gifted by FATHER.

And we must, by our greater free-willed loving agreement, find his shared understanding and belief among all loving souls. A brand-new conscious *and* physical state of perfect peace, joy, and love, lacking any new suffering, pain, or loss, can never exist with any level of hate (disagreement) with GOD on how to live. Such as eating a piece of forbidden fruit. And such lines can only be drawn by GOD, and not you or I. GOD says in Scripture, how can two walk-together, lest they agree. GOD further says to be of one mind (belief) and one accord (purpose). GOD'S purpose of eternal peaceful life among many, cannot exist for contrary divisive ways of life. Loving and hateful forces are projected in strengths from the slightest whisper to the loudest roar, and FATHER alone sees and draws the line where love turns to hate and becomes contrary to his eternal work. Now what love or hate do we each whisper or roar about, and at whom? Or from whom do we hear and receive the same? And then we must ask why? Materialist deny our sensing forceful creative soul exists, because they can't detect the empirical evidence of its existence. What about our forcefully created particle-wave based perceptions; these are empirical evidence. Yet without *any* evidence, science says our perceptions we do sense, are non-existent illusions. By the realm of Heaven existing, there must be an undetected ethereal spectrum to the periodic table that enables the *ethereal* qualities of a Heavenly environment required for its greater conscious/physical size and speed of life; being sensed and experienced by far larger, faster, more ethereal bodies and souls who require and interact with faster particles of light, in the way we interact with slower photons. The fallen do not know what forms or layers of evidence they see, or what to look for in defining reality. We emphatically declare up front, and no soul can show otherwise, that our created perceptions are all formed of higher or lesser particle wave properties, just like everything else we sense. And our consciously projected *intentional* quantized patterned forces of desire, are what create our perceptions within FATHER'S created physical spherical medium, or aether. All forceful patterned desires, projected into the physical spherical grid-like field, wave, or soup, of the aether, are the beginning of every physical complexity found above random chaos and disorder. To build an eternal perspective of reality, first and foremost, we must clearly understand what is the essence of force. Force is an indeterminable thing to fallen mankind. So, who among the mortals who walk this earth, can we ask to give us force's most profound definition? Who can declare for us its true first source, and true essence? Who can we trust to give us founding wisdom? When asked these questions, every scientist who can declare *all* that is known to mankind about force, must just stare back. Or they will reflexively skip the question, and speak from a prior existing force position, of reciting a consensus mantra of formulas used to calculate the observable *affects* and *effects* of force. There is no more to be shared by fallen mankind than measured pulling and pushing of attractions and repulsions; be they weak, strong, reaching near, or far. And beyond describing these affects and effects, forces true source and essence are unknown to mankind. I ask and trust no lesser soul for answers, that can only come from GOD, the source of all things. Above all things, we know there is only conscious desired, generated, projected patterned *intentional* force. Force is the original conscious projection of need, want, intent, and our will to exist and create our sensed reality we indwell. How can, and do *we* know such a thing? Again, FATHER shows *us*, and *we* see it in our own forceful desires and intents within our souls created three-dimensional visual and audio perceptions, and more. Prior to every awareness of our created perceptions, there is always our desired purpose of forceful intent to exist and know and be known. Simply stated, our every desire to exist is shown and seen to be forceful, projected, intentioned, patterned, and quite creative. The strong inherent forceful desire, or will of our soul to exist, is a true wonder to study. When medical science talks about the ongoing sustaining and perpetuating desire to exist, they say they see an enigmatic *willful* desire, or force, to live. And in a fallen environment, the intentional projected forces of binding love, and divisive hate, are warring aspects. Our soul, since our first desire, as we forcefully reached out to sense, and be entangled to, and know our physical environment, we have created our every perception, that bring our awareness. Wakeful perceptions are in large part, forcefully *created* substantive quantum echoes of the information we sense through the body. And by creating perceptions in our desired interpreted awareness, we build up our desired spiritual aspects of understanding and belief. Beliefs are our summation of all our interpreted information. Free-will, means we believe what we each want to believe. And a world filled with endless contrary beliefs, confirms, our free-will exists. We desire, interpret, and build up our understanding and belief, piece by piece. And being defined dimensional free-will-agents of awareness, we must create, interpret, and believe, by perceptions that fit within our limited 3D physical forceful reach. To define our three-dimensional space, or soul, we see our conscious sphere is centered by our desires, and filled with interpretations of understanding and beliefs. Our soul is our conscious shape, size, and reach of force, supporting all our spiritual aspects. And reversely, our soul exists up to and not beyond the edges of our desires, intents, and built awareness of understanding and belief. And all GOD has set externally, energetically, and physically, is there and shared with us all to be sensed to form our soul's interactions. Materialist see five aspects founding all reality, being force, energy, mass, time, and what *they* call an *illusion* of self- awareness. *We* see all souls set within GOD'S creation, and our perceptions we create and interpret, build up our beliefs. This world uses the word *force* to define all known attracting and repelling. But we see force far more deeply, seeing our soul's projected patterned desirous intent. *All* that we externally sense and more, rests and was formed within the immense reach and force of FATHER'S *firmer* perceptions of his awareness, belief, and understanding; the entire universe is his layered firmer perception. We create and perceive, just as GOD creates and perceives. We create layered *ethereal* perceptions, according to our minute size, reach, and force. And the Trinity create their *firmer* layered vast perceptions according to their immense size, reach, and force. This creative projected force process and outcome, occurs across vastly different scales and perspectives. All projected forceful purpose and intent, pulls, pushes, aligns, and creates all dimensions as it orchestrates all complexities, and all realms. Could the original forceful intents upon this universe have come from any source, other than a conscious force generator? Must not all creative force come from a true creative first source? Must force be newly generated and projected into existence? Or, do you believe that all force has always eternally existed? Does a great amount of force just exist and move around, creating the enigmatic complexity we see? We find zero physical evidence for that. So, force must originate and begin to exist by someone or something having a design in mind; being a supreme conscious soul's inherent projected forceful desired intention to exist. We see purposed force is everywhere it should and needs to be, *and* in its exact needed intended measure. We see no other reserves of force, just floating about, waiting to be used. Can you in any *other* way, describe the essence of conscious desire, without saying it is something that has the sole purpose of a willing wanting intent to cause something to happen. These words willing, wanting, and cause, which in essence describe the force of desire. Force is that which projects attraction or repulsion. Our every desired intent, is always to create or destroy; for something to come to us, or go from us. Nothing exists, that did not come into existence by forceful desirous intent, being the innate, inherent, generating attribute within us all, which has no prior existing force ascribed to it, or stored within it from any prior source. Do you see any prior source? Do not ascribe any small correlation as a cause in answering. In all that exists beyond all our souls, we see force being stored, transferred, and carried along in all created things; but never originating in them, or coming into a new existence by them. But we do see that our every new perception is created by a brand-new forceful desire, which must then be generated and projected. Yes, this can be argued against by vague secular desired beliefs, but never *successfully* argued against, or in any *slight* way disproven. We can never see or accept any other source but GOD for the external creation of anything. We see how only patterned intentional forceful desire, can order and build the complexities that form our multi-faceted 3D perceptions. And we see FATHER'S desired intending force, did ordain the once perfect energetic and physical complexities of the universe. Which are now weakened and stretched out by a rising rebellious momentum. Souls think they rail against GOD, but hurt only themselves. All ever-increasing disordered physical effects, come by billions of individual contrary forceful desires and intentions, going in differing individual directions, that they can never eternally sustained. FATHER'S displeasured withdrawing in turn, will allow the universe to soon dissolve. Most cannot see this, nor do they have any eternal forceful desire to do so. All who see eternally, gain a spiritual-*experiential*-qualia that can't be defined to those who don't see. Rebellion means two things, pushing GOD away, and increasing all disorder. In GOD'S UED model, mankind's standard model paradoxes, are all overcome, and a need to believe in hidden perturbations or variables to account for all their inconsistencies and missed predictions is gone. And all who cry and believe otherwise, can never prove otherwise. GOD'S model completely, and not partially, answers every *true* question you can possibly ask about physical and conscious reality. While mankind's model, inadequately argues against every true answer they have ever heard. GOD'S model is complete, mankind's is so incomplete. GOD'S model is immutable, and shows reality is orders of magnitude greater than fallen mankind can sense. Mankind's model is small, slow, weak, and in so many ways discredited. FATHER sacrificed out from himself and gifted many souls as angels and mankind in their conscious forceful orders and sizes to allow us all to desire, believe, choose, project, and experience our forces of love and agreement, or hateful disagreement towards GOD and each other. And the many, against all eternal-interest of love, joy, and peace, and all the good that was, and is now, and might ever be in their existence. They overwhelmingly follow each other into their individual endless insane demise. The fallen do not see how their temporary perspective of existence, revolves entirely around pleasing and experiencing the vile dying flesh. It is to by one's-self, or with another, or others, experience an empty temporary pleasure of having things put around, on, or into, or expelled from one's own, or another's flesh; such as cars, homes, clothing, dining, entertainment, sex, tattoos…etc. The flesh is exactly this crude and simple. As Eve did, and then Adam, they hear tempting lies of greater experience and pleasure, and run to inflict themselves with endless fruitless vile. There is only one possible cause for this, being the deep pride and envy of the flesh. These just must be a god over themselves; they cannot in their envy and pride, recognize and allow another to be *supreme* above them. They can allow smaller measures and appearances of greater adequacy in some who are around and beside them. But none so high above, as to rule over their believed reality itself. This is why those of a temporary perspective must see GOD as created in *their* image, and not the other way-round as commanded. They must have control to sense any temporary satisfaction they desire. Scripture says that pride and envy go before every downfall. Do we feel these elements battling within us against GOD and each other. It was only after I recognized and understood these elements, and their levels within me, that I was able to identify, fight against, and win these battles. To defeat an enemy requires we strengthen ourself, and know each enemy to the point of being able to foresee their attacks. If you truly enter this battle, you will quickly see it is a battle between your eternal spiritual desires for the next life, and your temporary desires for the flesh and this physical life. We find we are all forcefully governed, yet to differing degrees, enslaved to our acquired experiential desires that seek the sensations provided by this creation which come up against and stir our souls. Souls of a temporary perspective, seek out and *allow* these sensations, to become the far greater side of who they are. But as long as we live in the era of choice, the battle is winnable according to our understanding of it, and our growing desires towards GOD. Let's discuss the forceful weapons we can bring to this battle. Of course, love is the supreme force in all that exists, because it is the sole great desire of GOD. So, we simply adjoin and become one with the undefeatable, immutable, and undeniable love of GOD. The force of GOD'S love is freely given. So, the question becomes, how much do we truly want to receive and return this love? Do we want it all, or a lesser measure? Will some hold to a temporary perspective, and hit a point where their desire for more of GOD'S love fails? Yes. We see at each step of growth, we have traded a portion of our former reality, for a portion of GOD'S eternal reality. Some believers who read this must wonder, have they slowed or stopped at some past point of growth. GOD will never fail us; if we fail, it is always by our own core desires towards GOD. It has become apparent, that only the fewer want it all, as it comes down to how much of GOD and his *eternal* ways, we each desire. And without intimately knowing GOD, and holding the eternal perspective, it is all impossible. So, as we each come through study and prayer to intimately know and understand GOD more and more, and as our loving understanding for GOD grows, we see salvation is not a simple matter of just asking GOD. GOD cannot make us or anyone come to him in eternal love. If GOD could do that, he would have started there, and would not need to create this *costly* process to find his free-willed loving family. We are called to come together as one in belief and purpose to sustain life in an eternal loving peace, void of disorder. This requires that *no one* holds any core physical desires that contradict the purpose and needs of the whole. This requires learning and knowing what can hurt the coming new eternal reality GOD will soon create and sustain. We must submit in trust to his loving mentoring over us, and find all shared eternal good. In knowing your soul's desires are your mechanism of forceful creative projections; what are your waring desires. Evil desires, war against one's self-interest in unseen ways. Having the desires to both smoke and to quit, are waring desires. We must win all these wars. If we do not overcome all our waring desires, we can never sustain our eternal promise to GOD. Evil desires are born of a temporary perspective, that sees the world and the flesh in an individual way of desired existence. Please see how your belief and trusted understanding are so important. If you choose to create your own *individual* understanding, belief, and awareness of life, you find yourself in a most ridiculous position of waring against GOD'S created ways. You then become no more than a walking waring belief. Now we can also know ourselves more, by studying our dreams through the eternal perspective. Most believe that dreams are just dreams, not real, and not meaning anything. I humbly, firmly, and wholly disagree. Our dreams are our created perceptions within our soul, and our soul exists in our awareness of our dreams. Our every experience of awareness is real. So, it follows that all our deepest desires and fears, being who we are deep down, will rule over our sleep time perceptions, with less self-control. We see our wakeful awareness is expressed through our restraint, to not interact in an unaccepted form among others. So, why do we experience such uncontrolled levels of desires in our dreams? Our awareness in our sleep dreams is often experienced in having lesser control, less conviction, and surfacing our confusions as to life itself. But as we grow towards GOD, we notice our sleep dream reality becomes more controlled, like wakeful reality. For me, this has only been pronounced enough to notice after years of spiritual growth. This means my core desires are solidifying with my wakeful desires. And it grows more difficult to restrain who I am spiritually when awake, as former worldly façades fade, I become more outwardly true to GOD. I sense my eternal perspective growing, and temporary perspective fading; not by becoming one with myself, but one with GOD in belief, understanding, and ways of life. Yet I am still fully and always myself. But where I was once severely broken and lost, I am being gently and greatly healed. I sense my soul has become a far larger more focused presence through

GOD. Let me point out, that our shared spiritual beliefs about life itself, and our allowed physical preferences by GOD are not the same. For example, if we desire any of the ways of this world, that GOD says are not conducive with the *eternal* spiritual aspects of peace, joy, and love, such as feeling vain in one's attire and appearance, or if one feels they should have more say in their ways of life than GOD. These are prideful, envious, and destructive, to all eternal order. But, if in FATHER'S *provided* variety, one gently and gracefully prefers the fruit of a certain good tree, and the melody of a certain song, or certain good colored brilliance; and another soul gracefully and gently prefers a different fruit, song, and color. These desires of choice, are not among those divisive ways of life that hurt eternal love and the creation. We see FATHER gave all the wonders but one, in his great abundant Garden to Adam and Eve, to have and choose as theirs. FATHER needs, wants, and allows us to have many gentle preferences. We know we cannot all be identical in every aspect, and be agents of free-will. There are many allowed preferences in the eternal perspective; but we cannot cross that eternal sustainable loving line in how GOD has created and sustains all physical life. Science sees a Goldilocks Principle in our universe, that applies to all the processes of physical life. But they do not consider how it must firmly apply to *all* aspects of eternal life. In mankind's individual fallen desires, they see countless ways to live, that through their temporary perspective are seen as loving. Even as FATHER has said from his eternal perspective, they are divisive and unsustainable. I have found it uplifts me to give back to GOD, all control for defining and providing the ways of loving eternal life among many souls, which only he can create and sustain in an eternal way. And the burden of my every *need* has not only been met, my *eternal* provision has increased beyond comprehension. My every eternal physical need is now, and will always be met. I so treasure this loving peaceful presence, found in the growing oneness of *GOD'S* belief and ways. Before my spiritual resurrection, I held a certain desire to know and be known by GOD. But I held no intimate knowing understanding of who GOD was; so, I was still empty and void. But now, I have truly grown to know and see, *and* be known and seen by GOD; and in this my great desire is fulfilled. Many will sense this writing to be no more than arrogant superstitious verbiage. Some may feel I claim a spiritual superiority or authority above others. But *I* know my deference, shame, and fear before GOD, are strong and sincere. And I see my every accuser's prideful rebellion to the same. GOD has simply given us truth to live and share. Please hear this, as I do not seek to share *personal* beliefs with anyone; I have none to share. I only know and seek to share GOD'S one true belief. And GOD gave us a Scriptural command to share all we have with all who might hear. I obey this commandment to be performed in all love, fear, and humility, under GOD. But others by evil core desires, hear me claim Royal Authority and Judgment over them, which *I* neither claim nor have. JESUS *and* his Apostles, claimed they did not preach their own words and beliefs, but those given them from the above. Do any think JESUS or his Apostles were lying, were these just their own beliefs they made up. If any believes this lie, it is a rebellious perception of their own contrivance, and I have no responsibility for or to that. I do not mold myself according to any individual who is offended by the Bible. I will never deny one word of GOD, because another feels personal offence. FATHER and all his children are *continually* offended by what this world does and speaks; and not one liberal zombie cares the least bit that we are offended. The only claims of felt offence that matter to this world, are those of a self-righteous liberal consensus. The vile offence felt by FATHER'S Children in this world, means nothing to them. I see this world *deny* it is at war against the truth of GOD, as the fallen unite in hate against the same. They try to shame, rewrite, reinterpret, and take GOD'S words out of context to allow their ways. So, why should I expect any better treatment from them. If the truth of GOD is ridiculed and cursed by the lost; we who obey it will also be ridiculed and cursed. For speaking FATHER'S truth, JESUS was rejected, condemned, and tortured more than any other soul ever. The evil in this world screams out that those of the one true GOD, must not whisper his true beliefs; while all blasphemers, and all evil souls in this world scream their lying beliefs at us in every public forum. Many who want to stand with GOD are weak and intimidated by threats to stay silent. But GOD calls us to an even more difficult task than speaking out, he calls us to treasure the reprisal, abuse, and suffering we bear for sharing his words. This is *our* great wisdom in that we are called to lay our righteous sufferings before the throne of CHRIST as a gift. And our LORD sees we willingly suffered abuse in our love for him. Do you see that CHRIST loves you so much, because he willingly suffered immensely for you. Is that not loves greatest proof? Yes. So, in your suffering to stay with him, CHRIST will know you love him all the more. Once found, love in its small measures is affirmed by words and facial expressions. But loves great expressions are felt and proven in our sacrificial life of deeds and acts, even in our suffering death; just like the greatest possible sacrifice was endured by CHRIST. Fallen individuals do not experience the eternal truth found in our oneness with GOD. Every time I consider all that I have experienced and learned, and what I am becoming, I feel spiritual elation. We sense oneness with an eternal force of love, being something, we can never fully describe. It is an other-worldly sense, and a new-qualia of growing together as a part of who and what we are meant to be. Yet I know I am still me in every true way of being a free-willed entity. I do not see myself as being from the past, and in eternity I know the future never arrives. A new and formal form of the present moment, or pure Presentism, is all there is. We exist only and always in the now, as a spiritual eternal soul. We still must physically sense how we exist here in this world, yet we are not of, and do not belong here, as it is such an evil place. We treasure only every loving work that GOD might ask of us to help increase and sustain the Family and Kingdom, this being our eternal desire. We see this world is truly a divisive, destructive, and temporary illusion; and in all these things, we have no part, place, or desire. We see one path, that upon our freely desired choice, is created before us, just as we travel upon it. The many physical desires I once strongly sensed are now weak, many I have not felt for decades; yet I still sense my battles. Eternal love is revealed within us, only by the residing HOLYSPIRIT. We know we are moved by our core desires within, and these forces are our intentions and our chosen built-up hopes. Is GOD also moved by their forces of desire within? Of course. Can GOD'S core forces of desire evolve or devolve as ours can? No. We can change from our imperfect ways and desires to countless other imperfect ways and desires, or change to GOD'S defined perfect ways and desires. But FATHER'S fullness of perfect love is an *eternal* that can never change. Perfection and Eternal are synonymous; perfection is the purest state possible, being that which is without the slightest weakness or flaw. There are no levels of pure and full perfection. Pure and full perfection is that which nothing else in existence can alter, impair, or tempt. It is that which will remain perfectly intact for all eternity; it is FATHER. There will always be our perfect unfolding for us, but a true change to FATHER, could only be to diminish into the imperfections of death, suffering, and loss. We do not imply that FATHER ever evolved to be perfect. FATHER, in being first, vast, foremost, and all powerful, just is and has always been the definition of eternal perfection by being a supreme soul. This is just how all consciously determined reality works. FATHER'S eternal perfection and supremacy are in every way relative one to the other. Yet for fallen mankind, being born into, surrounded by, and drowning in conscious and physical imperfection, there is an escape. We must know that at our gifting and creation, we were of zero understanding or belief. We were in all essence, all gifted blank conscious probability potential equations of desire and choice. And *our* eternal perfection, can only come to exist in our emulating CHRIST'S belief and love that can sustain us all together. We are created insufficient in every eternal self-sustaining way. And in our finding GOD'S perfect way, we find our perfect*ing* always remains before us in loving and following that which *is* perfect and eternal. I need not tell anyone, that thankfully we have an eternal loving FATHER above. For if we had an evil FATHER above, like Satan, our love for him would not matter, and there is no end to the hell we all might forever endure. But we have proof FATHER is loving, as is our LORD, and HOLYSPIRIT. This is also so of all the *Holy* Angels and *Holy* Saints. But Satan and all rebellious angels and mankind, are not desirous of being perfect or loving towards GOD, and are called devils. All souls not following

GOD'S loving ways of belief and understanding, are devils to GOD. In GOD'S eyes, there are only the highest to the lowest of good and evil souls. It is everyone's freely chosen and exercised conscious desire that defines which they are. It is all about GOD'S search for the final separation of two free-will, freely chosen, opposing forces; being a oneness of pure creativity, and a divisiveness of pure destruction. These two desires, define and divide these two ways. Does this allow a middle ground of being neither, or both? No. Has science or theology ever seen or found any non-relative unaccountable voided area, existing between any two opposing forces, or between any two created realities. We see only our ever-expanding moment of searching, choosing, and finding our greatest love; then comes our sealing by GOD in our desired eternal choice. But during *our* stranger moments of choosing, changing, and growing, we are found to be on one side or the other. And only during our short moment in this world of choosing, can our changing desires move us back and forth. But after our moment of choosing our final path, we will be set in the immutable eternal and forever. Now our love for GOD is not always equal to our understanding of GOD. And our love for GOD does not come by our understanding of GOD alone. I tell you there have been souls who find their pure obedience to GOD, yet they may not understand all his creative ways. Some souls just love GOD so much, they instinctively study and live in GOD'S ways, in a perspective far above their level of understanding reality. Now I must not envy these. But I recognize these came more by great love and adoration; where I sense my original love has now grown by my understanding. I loved GOD before, but I love him more now, yet I can't equate every cause for this difference. This truth can be taught to those who love without knowing, adding even more to their great love for GOD. But if my love gained through understanding GOD'S reality, should still be an insufficient love, my knowledge and understanding of GOD and reality will not matter. The understanding of GOD can be shared with us, but a love *for* GOD, by GOD, cannot be instilled in us. Our understanding is truly important, as it enlightens and strengthens our love; but a naïve trusting love has a beautiful rarer childlike aspect. Some just love GOD more, and some loves grow more in learning of GOD. A great obedience for love alone, may matter more, above a great obedience for loving understanding. But our obedience for both reasons, must be our greatest measure. I believe those who love and obey GOD as much or more than me, without understanding GOD'S reality, are more precious. Yet I wish they knew GOD'S reality, so they could love GOD even more. There are forces here, and forces there, the ruling desires of conscious force are everywhere. Once we accept that our soul is an eternal projector of newly generated 3D patterned forceful desires that create our perceptions, by aligning their wave particle properties. Only then can we begin to see the created nature of GOD'S reality. We must recognize that all creative forces, come only from within all souls. We must also accept that the ways of eternal life were created and set by FATHER before this reality, or we, even existed. If any thinks the ways of GOD'S coming Kingdom are not yet set, and that these ways may come by a democratic process; this simply cannot be, as it could never be perfect, or endure. Most believers walk in confusion, thinking the boundaries, properties, and ways of GOD'S Kingdom, are established according to the desires of a forgiven yet flawed family. The only way to see a heaven filled with a flawed family, is to see a flawed and temporary heaven. Do you see Heaven around us now, *or* do you see many souls, having free-will and wanting different ways of life, is exactly equal to the division, violence, loss, and death we now see. I still sense certain fading residual desires for this world; but I recognize and understand them. And if an evil desire rears its head, I know it for what it is; being against the Kingdom of GOD, so, I can better resist and win in CHRIST'S name. We are in both an external and internal battle; choose GOD'S ways and strive to stay pure. This is part of our eternal perspective. This is the only way to eternally know who and what you are. This moment, is our time of choosing who we will eternally be; yet there comes another long moment of their choice for some in CHRIST'S Millennial Reign. This millennial moment is for those who in any way, are not set in their final choice. By GOD'S perfection, these warrant their moment of trying. But the vast majority who die, are already separated and sealed against that day. It can be hard for some to understand, that at the close of this third dispensation, there will be some who died, and some yet living, who did not make their final choice beyond turning around, And the perfect justice by GOD must be served. In this Millennium, most souls will still lose their love and turn against JESUS, as most of mankind always do. 2nd Peter 3:15-16, "*EVEN AS OUR BELOVED BROTHER PAUL ALSO ACCORDING TO THE WISDOM GIVEN UNTO HIM HATH WRITTEN UNTO YOU; AS ALSO IN ALL HIS EPISTLES, SPEAKING IN THEM OF THESE THINGS, IN WHICH ARE SOME THINGS HARD TO BE UNDERSTOOD, WHICH THEY*

THAT ARE UNLEARNED AND UNSTABLE WREST, AS THEY DO ALSO THE OTHER SCRIPTURES, UNTO THEIR OWN DESTRUCTION". This Scripture speaks of many religious souls seeing their self as good, by accepting their own definitions of good and evil. Most souls hear their Scriptural interpretation, and never hear HOLYSPIRIT'S interpretation. Beyond dispute, the *current* spiritual nature of this world is evil. But this physical reality ultimately obeys GOD'S desires. FATHER is entangled to all physical reality, and controls it in an eternal purposed way. FATHER has always existed, and we cannot define what always means in a linear way. FATHER is spherically eternal, not linear; and he is unchanging, perfect, and loving, having omniscience. The only way to change from perfection and omniscience, is through the impossibility of becoming less than omniscient, and less supreme to a point of being vulnerable to another external coercive conscious soul. We see in Scripture, that the unknown quantities of the qualities that define FATHER were before the beginning of all that is now. And FATHER was utterly alone as a solitary dreaming entity. Isaiah 44:6, last sentence, "*I AM THE FIRST, AND THE LAST; AND BESIDE ME THERE IS NO GOD*". Isaiah 44:8, last part, "*IS THERE A GOD BESIDE ME? YEA, THERE IS NO GOD, I*

KNOW NOT ANY." Isaiah 45:5 "*I AM THE LORD, AND THERE IS NONE ELSE, THERE IS NO GOD BESIDE ME.*" This speaks of a search, made by the lonely supreme fullness of FATHER, *before* all his gifted conscious divisions, and his external created aether, Heaven, and universe. Can we try to imagine being the only soul to exist? In my weakness it sounds terrifying. Only a perfect loving FATHER could ever overcome such a horror. And now we know pure loneliness, is part of FATHER'S eternal omniscience. Even JESUS and HOLYSPIRIT, never knew this loneliness; though CHRIST came close on the cross, as FATHER and all Heaven looked away. All created certainty, came by the singularities that form the body of FATHER. And all souls below FATHER are lesser creative souls; or lesser to far lesser creative Gods to be precise. As FATHER said in PSALMS 82:5-7, "*THEY KNOW NOT, NEITHER WILL THEY UNDERSTAND; THEY WALK ON IN DARKNESS: ALL THE FOUNDATIONS OF THE EARTH ARE OUT OF COURSE. I HAVE SAID, <u>YE ARE GODS</u>; AND ALL OF YOU <u>ARE CHILDREN OF THE MOST-HIGH</u>. BUT YE SHALL DIE LIKE MEN, AND FALL LIKE ONE OF THE PRINCES.*" In John 10:34-36, "*JESUS ANSWERED THEM, IS IT NOT WRITTEN IN YOUR LAW, <u>I SAID, YE ARE GODS</u>? IF <u>HE</u>*

<u>CALLED</u> <u>THEM GODS</u>, UNTO WHOM THE WORD OF GOD CAME, AND THE SCRIPTURE CANNOT BE BROKEN; SAY YE OF HIM, WHOM THE FATHER HATH SANCTIFIED, AND SENT INTO THE WORLD, THOU BLASPHEMEST; BECAUSE I SAID, I AM THE SON OF GOD? IF I DO NOT THE WORKS OF MY FATHER, BELIEVE ME NOT." Yes, the soul of JESUS, "the beginning of the creation of GOD" and "the firstborn of every creature" most assuredly came forth out from FATHER'S soul. And JESUS speaks of this in Proverbs 8:13-36, "*THE FEAR OF THE LORD IS TO HATE EVIL: PRIDE, AND ARROGANCY, AND THE EVIL WAY, AND THE FROWARD MOUTH, DO I HATE. COUNSEL IS MINE, AND SOUND WISDOM: I AM UNDERSTANDING; I HAVE STRENGTH. BY ME KINGS REIGN, AND PRINCES DECREE JUSTICE. BY ME PRINCES RULE, AND NOBLES, EVEN ALL THE JUDGES OF THE EARTH. I LOVE THEM THAT LOVE ME; AND THOSE THAT SEEK ME EARLY SHALL FIND ME. RICHES AND HONOUR ARE WITH ME; YEA, DURABLE RICHES AND RIGHTEOUSNESS. MY FRUIT IS BETTER THAN GOLD, YEA, THAN FINE GOLD; AND MY REVENUE THAN CHOICE SILVER. I LEAD IN THE WAY OF*

RIGHTEOUSNESS, IN THE MIDST OF THE PATHS OF JUDGMENT: THAT I MAY CAUSE THOSE THAT LOVE ME TO INHERIT SUBSTANCE; AND I WILL FILL THEIR TREASURES. THE LORD POSSESSED ME IN THE BEGINNING OF HIS WAY, BEFORE HIS WORKS OF OLD. I WAS SET UP FROM EVERLASTING, FROM THE BEGINNING, OR EVER THE EARTH WAS. WHEN THERE WERE NO DEPTHS, I WAS BROUGHT FORTH; WHEN THERE WERE NO FOUNTAINS ABOUNDING WITH WATER. BEFORE THE MOUNTAINS WERE SETTLED, BEFORE THE HILLS, WAS I BROUGHT FORTH: WHILE AS YET HE HAD NOT MADE THE EARTH, NOR THE FIELDS, NOR THE HIGHEST PART OF THE DUST OF THE WORLD. WHEN HE PREPARED THE HEAVENS, I WAS THERE: WHEN HE SET A COMPASS UPON THE FACE OF THE DEPTH: WHEN HE ESTABLISHED THE CLOUDS ABOVE: WHEN HE STRENGTHENED THE FOUNTAINS OF THE DEEP: WHEN HE GAVE TO THE SEA HIS DECREE, THAT THE WATERS SHOULD NOT PASS HIS COMMANDMENT: WHEN HE APPOINTED THE FOUNDATIONS OF THE EARTH: THEN I WAS BY HIM, AS ONE BROUGHT UP WITH HIM: AND I WAS DAILY HIS DELIGHT, REJOICING ALWAYS BEFORE

HIM; REJOICING IN THE HABITABLE PART OF HIS EARTH; AND MY DELIGHTS WERE WITH THE SONS OF MEN. NOW THEREFORE HEARKEN UNTO ME, O YE CHILDREN: FOR BLESSED ARE THEY THAT KEEP MY WAYS. HEAR INSTRUCTION, AND BE WISE, AND REFUSE IT NOT. BLESSED IS THE MAN THAT HEARETH ME, WATCHING DAILY AT MY GATES, WAITING AT THE POSTS OF MY DOORS. FOR WHOSO FINDETH ME FINDETH LIFE, AND SHALL OBTAIN FAVOUR OF THE LORD. BUT HE THAT SINNETH AGAINST ME WRONGETH HIS OWN SOUL: ALL THEY THAT HATE ME LOVE DEATH." How shall we come to JESUS. Well, it says in Colossians, 1, 11-29, "*STRENGTHENED WITH ALL MIGHT, ACCORDING TO HIS GLORIOUS POWER, UNTO ALL PATIENCE AND LONGSUFFERING WITH JOYFULNESS; GIVING THANKS UNTO THE FATHER, WHICH HATH MADE US MEET TO BE PARTAKERS OF THE INHERITANCE OF THE SAINTS IN LIGHT: WHO HATH DELIVERED US FROM THE POWER OF DARKNESS, AND HATH TRANSLATED US INTO THE KINGDOM OF HIS DEAR SON: IN WHOM WE HAVE REDEMPTION THROUGH HIS BLOOD, EVEN THE FORGIVENESS OF SINS: WHO IS THE IMAGE OF THE*

INVISIBLE GOD, THE FIRSTBORN OF EVERY CREATURE: FOR BY HIM WERE ALL THINGS CREATED, THAT ARE IN HEAVEN, AND THAT ARE IN EARTH, VISIBLE AND INVISIBLE, WHETHER THEY BE THRONES, OR DOMINIONS, OR PRINCIPALITIES, OR POWERS: ALL THINGS WERE CREATED BY HIM, AND FOR HIM: AND HE IS BEFORE ALL THINGS, AND BY HIM ALL THINGS CONSIST. AND HE IS THE HEAD OF THE BODY, THE CHURCH: WHO IS THE BEGINNING, THE FIRSTBORN FROM THE DEAD; THAT IN ALL THINGS HE MIGHT HAVE THE PREEMINENCE. FOR IT PLEASED THE FATHER THAT IN HIM SHOULD ALL FULLNESS DWELL; AND, HAVING MADE PEACE THROUGH THE BLOOD OF HIS CROSS, BY HIM TO RECONCILE ALL THINGS UNTO HIMSELF; BY HIM, I SAY, WHETHER THEY BE THINGS IN EARTH, OR THINGS IN HEAVEN. AND YOU, THAT WERE SOMETIME ALIENATED AND ENEMIES IN YOUR MIND BY WICKED WORKS, YET NOW HATH HE RECONCILED IN THE BODY OF HIS FLESH THROUGH DEATH, TO PRESENT YOU HOLY AND UNBLAMEABLE AND UNREPROVEABLE IN HIS SIGHT: IF YE CONTINUE IN THE FAITH GROUNDED AND SETTLED, AND BE NOT

MOVED AWAY FROM THE HOPE OF THE GOSPEL, WHICH YE HAVE HEARD, AND WHICH WAS PREACHED TO EVERY CREATURE WHICH IS UNDER HEAVEN; WHEREOF I PAUL AM MADE A MINISTER; WHO NOW REJOICE IN MY SUFFERINGS FOR YOU, AND FILL UP THAT WHICH IS BEHIND OF THE AFFLICTIONS OF CHRIST IN MY FLESH FOR HIS BODY'S SAKE, WHICH IS THE CHURCH: WHEREOF I AM MADE A MINISTER, ACCORDING TO THE DISPENSATION OF GOD WHICH IS GIVEN TO ME FOR YOU, TO FULFIL THE WORD OF GOD; EVEN THE MYSTERY WHICH HATH BEEN HID FROM AGES AND FROM GENERATIONS, BUT NOW IS MADE MANIFEST TO HIS SAINTS: TO WHOM GOD WOULD MAKE KNOWN WHAT IS THE RICHES OF THE GLORY OF THIS MYSTERY AMONG THE GENTILES; WHICH IS CHRIST IN YOU, THE HOPE OF GLORY: WHOM WE PREACH, WARNING EVERY MAN, AND TEACHING EVERY MAN IN ALL WISDOM; THAT WE MAY PRESENT EVERY MAN PERFECT IN CHRIST JESUS: WHEREUNTO I ALSO LABOUR, STRIVING ACCORDING TO HIS WORKING, WHICH WORKETH IN ME MIGHTILY." JESUS says in Revelation 3:11-16, "*BEHOLD, I COME QUICKLY: HOLD*

THAT FAST WHICH THOU HAST, THAT NO MAN TAKE THY CROWN. HIM THAT OVERCOMETH WILL I MAKE A PILLAR IN THE TEMPLE OF MY GOD, AND HE SHALL GO NO MORE OUT: AND I WILL WRITE UPON HIM THE NAME OF MY GOD, AND THE NAME OF THE CITY OF MY GOD, WHICH IS NEW JERUSALEM, WHICH COMETH DOWN OUT OF HEAVEN FROM MY GOD: AND I WILL WRITE UPON HIM MY NEW NAME. HE THAT HATH AN EAR, LET HIM HEAR WHAT THE SPIRIT SAITH UNTO THE CHURCHES. AND UNTO THE ANGEL OF THE CHURCH OF THE LAODICEANS WRITE; THESE THINGS SAITH THE AMEN, THE FAITHFUL AND TRUE WITNESS, THE BEGINNING OF THE CREATION OF GOD; I KNOW THY WORKS, THAT THOU ART NEITHER COLD NOR HOT: I WOULD THOU WERT COLD OR HOT. SO THEN BECAUSE THOU ART LUKEWARM, AND NEITHER COLD NOR HOT, I WILL SPUE THEE OUT OF MY MOUTH." Now fallen mankind says they see different external mysterious unsourced forces. But those forces, are sourced, and they come only by GOD'S reasoned conscious loving forceful intent. Romans 13:1, "*LET EVERY SOUL BE SUBJECT UNTO THE HIGHER POWERS. FOR THERE IS NO POWER BUT OF GOD: THE POWERS THAT BE ARE*

ORDAINED OF GOD." If you can't understand and accept, that *one* great immense conscious soul, is the source of all external universal force, far beyond that of *our* lesser souls. Then the four forceful distinctions of consistency you see set over FATHER'S firmer external perceptions, must take on an improper appearance. For example; FATHER, by the fastest light particles, sees his entire dynamic universe, exactly whcrc, and as it is now, in all its movements each moment. From the spin upon every particle to the largest firm realities, FATHER sees it all, but mankind cannot. So, they turn to project their gifted force *against* FATHER who gifted it to them. Imagine giving someone a beautiful eternal gift out of love, being the forcefulness to exist, and they turn to try to destroy you and your gift. Imagine sacrificing out of yourself, and giving billions of these gifts of life, and en-masse, most turn on you in rejection and contempt. I have heard people make a self-aggrandizing sacrilegious joke about themselves; saying it's not easy being God. I see from my lowly understanding, that it is *currently* much harder being the one true GOD, than for all souls combined to be themselves. But GODS eternal rest and just reward soon come. If I see cruelty in this world, it is then a cruel world. If I see death in the world, it is then a dying world. If I see hate in the world, it is a hateful place. But we also see love, and great patient works, and the long sufferings efforts of GOD to find and bring us the wondrous gift of eternal life. Now mankind did not fall from every perfect word of GOD in one fell swoop. They fell from one perfect word first, and then continued to fall further from every other perfect word. So, all who are restored to GOD, come back only by obeying every word, one by one. Mankind's disobedience is the cause of all painful loss, and to return on high, requires our renewed obedience to every loving word. So, I chastise and warn *myself* and all, against any misuse or abuse of GOD, or their names, or their wondrous gifts. Do not take their Holy Names upon your lips in vain. We know GOD will have their eternal sustaining labors in the next New-Heaven; and in our love, we must keep it at GOD'S desired chosen level. Please see GOD has accepted and taken up an eternal never-ending labor to sustain a New Heaven and Earth filled with around 300,000,000 angelic bodies, which GOD must also sustain. And GOD must sustain a great lake of fire with around 70 billion eternal bodies connected to fallen souls. Imagine taking on this eternal responsibility of work, just to have and sustain the eternal love of Family. And we too, must take on our eternal faithful responsible ways of eternal life; adjoining our small force of love to the good loving forces of GOD. We must dwell faithful within the eternal wonders prepared for all who truly love and become one within the creations of GOD. I sense the balance of my desires and oneness, and since 1974, I have learned to fight my evil side. My soul was once so far off and so contrary to GOD, and I still have my unknown distance to go. My desires to align with GOD'S created ways grows, and my *eternal* love and perspective are increased. But certain weaker remaining desires for physical temporary things are still fading, so, my eternal uncertainty remains. I can know in a certain moment I am forgiven and new; but Judas did too. Our current winning closeness to GOD, brings us blessed assurance, but not eternal assurance; not yet. We gain greater understanding of good and evil, in *exercising* the good, but we need only see it, to understand evil. We cannot do this, holding to the confused ways of this rapidly falling world. I plead with all to seek the experiential awareness of GOD'S sensed eternal qualia, that can't be described by words alone. The eternal perspective brings new meaning to the term qualia. I promise the only things you have to lose in this, are your unrighteous hate and pain, and your hell and second death, all being part of a temporary perspective. Yet, many in pride, envy, and hateful delusion, worship such evil things. This appears evermore paradoxical to us with our growth. They hold to an insane understanding we no-longer, and hopefully never again comprehend. Their contrary understanding, bring forms of false conscious qualia that come with a temporary perspective. I know a temporary-sense exists I no-longer experience as a primary perspective. As we lose our temporary perspective, our awareness of its spiritual qualia fades. It is like forgetting what certain foods taste like, and being replaced by others you now love. This is a bit strange, and if I turn back *away* from GOD, and lose my eternal perspective completely; which sounds insane to me. This Perspective Principle states I must quickly lose my experiential sense and remembrance of these eternal quale or qualia; and again, acquire my former temporary sense. I have experienced both forms of contrary spiritual qualia; but I have not, and cannot experience or remember them both at the same time. Are there more experiential qualia to come with growth, I expect much more. Many say it's bragging and rude to claim I have, or sense, something another might not. This may be *true* of the great temporary physical riches of this world, which are set beyond the reach of most by chance and their choices. But this is a spiritual gift, set only beyond the reach of those who do not want it enough to go to it, or do not believe it exists. So only an evil soul could proclaim me rude or bragging by declaring I have searched for and received something from GOD, that they and all can have. And if I failed to disclose this truth, these would be the souls on judgment day who accuse me of hiding and not sharing this understanding with them. With such narcissistic souls, I can never converse without it turning to insane arguments and accusations born of their confused small slow physical perspective. Scriptures say to speak truth, move on, and not waste our efforts on such souls. But I have against GOD'S instruction, tried too many times with certain souls. This is a zeal of false hope, that can over-ride my understanding. But more and more, I follow Scripture on this. Now please know that *every* strange beautiful physical pattern, found in GOD'S creation, was put there by GOD. There are no patterns in this reality that were not created; no sustained repeating patterns ever evolved or emerged. There are also now weakened and diminished remnants of GOD'S patterns, but none that were not first created perfect by GOD. Now the force of GOD'S creative mind, works on unimagined layers and speeds, far beyond the fallen sense of time. FATHER sees reality from all omniscient omnipresent superpositions, as his forceful patterns of sustaining guidance come down through his many creative mechanisms. And in the quantum ethereal realm of all souls, these terms conceptual and superposition, do not mean physically non-existent. Our every *thought* has its created physical half in our substantive particle-wave-based perception of that thought. There exists no physical pattern, that had a beginning, *not* created by a soul. All firm external or ethereal internal creations, are but two different observed strengths and forms of the one same act of conscious forceful creation. Every conscious patterned concept can be brought to physically exist by FATHER, even physical life. There are at play here, *our* minute ethereal created perceptions, within *GOD'S* vast immense created firmer perceptions. Now the probability wave aspect, described by mankind in quantum measurements, is in reality a growing phase of building up all patterned particle reality. All energetic waves, are actually a vast soup of consciously projected forceful moving points and particles, too small to be known by mankind. But as the building formative process and layers of all desired particle creations, build physically large enough, they can be detected by mankind. So, mankind's moment of conscious certainty, comes at the moment each complexity grows large enough to be perceived in their slow small physical awareness. In this conscious forceful process, from start to the finish, we see not one physical control, cause, or influence. And *all* lesser souls creative force patterns of desire, interact in an affirming or divisive ways with FATHER'S creation. Now again, GOD uses many conceptual sets or forceful creative conscious tools, all seen from his super-position of seeing every point and angle at once, being omnipresent. These mechanisms focus GOD'S many sets of desired forceful functions, down into all the minute levels to orchestrate physical life. But mankind does not see these pattern sets the way GOD does. Mankind from a slow small temporary perspective, sees all patterns start from the number one; seeing a meaningless entangled rising beauty. But not seeing their downward true purpose. From the eternal perspective, all repeating patterns are seen as narrowing funnels of great forceful guiding complexity, projected down towards one entangled purpose. *One* complex purpose is the final goal, not a numerical starting point. These patterns of force, allow GOD to project his complex willful work repeatedly, without coming down to one by one, perceive each and every minute forceful action of life. Consider how one *complex* patterned cell of life has many forced functions, working under one top-down pattern of forceful control, to being entangled as one lively system. And a human body, has many complex systems, that work together in one forced goal of being entangled as one system of complex life. FATHER'S patterned force, bring so many orders of guidance down into one purpose, which could never be formed in chaos, or start by adding one to one, out of trillions of trillions of ones of anything. All GOD'S complex patterns are created towards a complex final purpose. Why would a soul who believes GOD created all things, see these patterns as starting from an evolutionary point of one, and having no observable purpose. GOD creates nothing without purpose. Why see orchestrated beauty, in a meaningless ugly purely physical way, and not the other way-round. We see patterns of guidance for life, must be forced and funneled down for a purpose from above. Which is the true top-down consciously projected design of all information, and every complexity it forms and guides. The most-complex to most-simple physical patterns are purposed information, funneled down by FATHER'S creative desires. These patterned forces that guide life, are seen only through the eternal perspective. This, as materialists declare without any recognized logic or proof, that all enigmatic patterned complexities form from one point to two, and so-on, guided by disorder, chaos, and chance. All without any consciously intended purpose. And never forget that *as* they say they see no need for GOD, they start every theory upon a great unexplainable miracle. Without their founding miracles they have nothing. Their physical miracles are their godless faith and self-deceit, their bait and switch, and their unrecognized self-inflicted defeat. Only upon accepting their presupposed fantasy, can they *then* string together their loose-fitting false concoctions, as their model and theories add up less and less. We must see that complexity in everything including our perceptions is always projected as one entangled force pattern, creating complexity from a knowing source, in formed intent. Open your eyes to see this is perfectly logical and feasible. Accept that the universe's vast complexities, profoundly appear to have the purpose and intent of supporting physical life and enabling our conscious choices of belief. This could never happen by chance; even if started by a physical miracle. For me, this declares our immense, consciously creative GOD, to be a *certain and immutable necessity*. Can materialist see their undefined construct of time itself, being reverse engineered back into a singularity with the four known forces, which all refuse to cooperate. If any can't to see this, they can't see what we convey here. And all who do see it, are staring into the eternal perspective; which is only seen by GOD above, sharing it with us. At our first sensing GOD'S understanding within our soul, we then and there, have had at minimum, an actual peripheral connection to the projecting mind of GOD. In our desire, we open up to be literally touched by the shared mind of GOD. These understandings flowing down from the eternal source of all reality, can at first be hard to grasp in completeness, it is an immense spatial reality to comprehend all at once. But remember, our difficulty is based upon our pureness, resistance, and intent; and our growing in GOD'S *eternal* purpose and intent can feel slow at times. If this makes no strong sense to you, I hope you fast, study, and pray on it many times in hope that it will. My prayers are with all who strive for GOD'S gift of truth. Why do I not pray for *all* mankind; because from the eternal perspective, I know it can never be. I only pray for what can be. Life is so far beyond mankind's understanding, as all its complex aspects and attributes hang entangled together by GOD'S rich forces. Remember, this was all perfect in its original creation, and what we now see are its fading remnants. All laws of nature, were at creation, perfectly born of strong prime values, and many top-down patterned mechanisms. But after souls fell into rebellion, they began to one first, and complexity second, seeing reality from the bottom up, or inside out. The truth was severed in them by their reversed perspective and desires. FATHER is nearly ready to cease all sustaining mechanisms from above. And after CHRIST'S Millennial reign, it will all be suddenly lost and dissolve. If mankind could see the layers of forced patterns, mechanisms, and sets, that form the entire creation's original information, they would see how all laws and constants have devolved. And see it can all be reverse engineered back to the original perfect creation, around twelve thousand years ago. Even in its approaching dissolution, we see a remnant of near perfect tuning to the universe. But FATHER'S desires for this fallen creation have faded and approach a sudden event mankind does not see coming. Why? Because they refuse to see his conscious controls upon it all. After the millennial reign, all top-down forced entanglement will be completely cut. It will all suddenly fission and dissolve back into the single points of the aether, as if it never were. This universe, in little more than 1000 years from now, will meet its foreseen, foregone, foretold, conclusion. There is not a single thing that any, or all souls under GOD can do to stop this. But the end of *this* creation, signals the creation of the *next,* being sealed and eternal. Please see how complex quantized forced patterns, equals the complex pulling and pushing of guidance that creates, or destroys. Our soul's projected patterned force, is our intentioned information, meant to align the most basic quantum aspects to mirror our purposed desire. The science of mankind, sees themselves atop all they can sense; but their slow small perspective restricts and prohibits their ability to see or comprehend the greater and lesser conscious and physical realities above and below and beside them. Their beliefs tie their spiritual feet together, they tie their spiritual hands behind their backs, they plug their spiritual ears and put spiritual blinders upon their eyes. And then in hubris, they say through their very limited physical senses, and small slow temporary perspective, they understand and can define for *us*, all that exists. What words does science use, to describe the essence of force: mystical, mysterious, other-worldly, and enigmatic, are a few. But *we* would add all antithetical descriptive pairs, such as creative force or destructive force, truthful or deceptive force, responsible or irresponsible force, faithful or unfaithful force, loving or hateful force, and every-one across all paired spectrums. These are all conscious forces of intention, that define our perspectives, beliefs, and perceptions; and they also act to strengthen or destroy FATHER'S created realities. We're all conscious souls, desiring and choosing to project aligning or contrary, good or evil, forces of intent and purpose. I study and strive to correct all my forces. The projected forces of GOD and us, aligning, converging, and strengthened in all ways, is the only eternally sustainable way for love and life to endure in a reality requiring a perfect eternal physical half. And by GOD, the physical is provided to us in our need; so, we should lovingly, responsibly, and faithfully, choose to align our souls to every good force of life. Do you understand who and what you are, and what you forcefully project, choose, say, do, and create? Do you truly know these things? Or, are you flailing about, blind to all that is above and beyond your physical line of sight and sensing? Most souls are enamored with, and focused on the little temporary fading physical pleasures they might enjoy before it all ends. Most souls are now more animal-like, than Godlike. If you are in such a state, I exhort *you* to untie yourself from such anchors; open your ears and eyes and look up and up to where the one eternal truth dwells. Find not who you are right now, but the forceful wonder and part of GOD you were created in hope and love to become. All projected force is a true intelligible intended aspect. If the entire world in one loud voice said to me that I am wrong, and that force is just an unknown fact. I could not be moved in any tiny measure from the truth I see and know from the highest authority. I've experienced the inalterable highest aspects of love, I've felt the immutable highest orders of love. I see the highest fastest lights in their eternal contrast to the lowest slowest darkness. How could I be called and dissuaded again by the aspects of a lesser lower darkness, to leave off from knowing the greatest highest brightness. Yet I still live in my own great caution and trepidation of all the coming great evil deception being prepared against us upon the earth. By every pronounced darkness that grows in this world, I know I am fallible if I am *ever* found outside of being one with GOD. The world wants and tempts us all in its many ways to forget GOD, for but just a little moment, and to sin but just a little bit. I have surely not yet seen nor overcome every single dark deception to come; but I look ahead and to the sides, behind, above, and below, for they will always come from all directions until my last moment in this world. In Scripture, we read temptations came fast, strong, relentlessly, and disrespectfully against the only begotten SON who prevailed and is perfected. Do we warrant less temptation? If we see evil approaching through our eternal perspective, it is obvious and has no powers of deception that we do not first surrender. I warn you of the growing darkness of spiritual deception. It is strong and getting stronger, fierce and growing fiercer, it is desperate and growing more so. It is angry, deceptive, vicious, unrelenting, and deadly. Run with all you are, and be folded into GOD'S eternal love, to be safe and saved. Let me share an experience from 1974 that occurred soon after my own rescue. I was a young man of 20, out walking and witnessing. And upon passing by and asking a couple out in their front yard if they had an interest in GOD, I was invited into this young couple's home to talk. Thcsc young newlyweds were open and loving, yet unknowing of GOD'S ways. This was a chance meeting as they did not know me; or the dark soul who knocked on their door a little later as we spoke. I was young and clean, but with long hair, wearing flip flops and bib-overalls, carrying my Bible. Not a normal visual of a Preacher, I was what was back then, called a JESUS freak. I spoke loving from my heart with no prepared thoughts. But the dark soul who knocked at the door within the hour was polished, wearing black slacks, a white long sleeve shirt and tie. He had the look of a worldly professional authority. Now it is relevant, when I say that this was in the very hot South Florida summer, and this couple had no A/C, only a fan that blew warm air. At the knock on the door, they got up and let this dark soul in. Now I had never heard of the phenomenon before, so, it was not a psychological suggestive affect, that when a man about 6'4, with black immaculately trimmed set hair walked into the kitchen where I was sitting. And the temperature in the room to me, but not the couple, grew *noticeably* cold. He quickly spoke of a new church he had started, designed for young couples. I felt a sudden change in these young souls as his deceptive hook was dangled in front of their unknowing eyes; and they quickly and strongly took it. Sadly, I had just lovingly for about an hour, peaked their desires towards GOD and set a table for evil to walk in and steal them away. They did not understand that the truth cannot be tailored to suit or help people of different ages or status. They did not know all souls must come and grow together in GOD'S ways, and not the ways of young worldly couples. After he left, I told them he was misleading them, but they were turned, just that easy. I was sad at how their simple core desires to find and worship GOD among other young couples did separate and take them down a weakened path. How and by who we are first introduced to GOD in our weakness, is quite often very important. If a weak worldly preacher sees GOD through a temporary perspective, they preach GOD from that same wrong perspective. I do hope that couple found their way out. Now from a temporary perspective of love, it is most-often not seen, how one small choice could matter or affect an over-all final outcome. But from the eternal perspective, the greatest to the least of all dangers are seen as deadly. In the eternal perspective, a great eternal good, or evil, meaning appears within every sentence, song, sentiment, hope, intent, and in every love. And failing to sense this, and falling into a blind acceptance of endless *tiny* dangers is how most souls are defeated. Most souls could never succumb to the greater evil of murder, as these are such strong and easily sensed evils. But through desensitizing and acceptance of *ungodly* tolerance, souls grow accustomed and numb, and even desirous of tiny evils. Lesser evils, through core desires for the same, are seen as being okay or not evil. Like what Eve and then Adam thought about a seemingly harmless piece of fruit. They disbelieved, and forgot what GOD said was good, and for themselves decided it was good. See all that is defined by GOD as evil; then see how mankind, having no authority, keeps moving things one by one from the evil side to the good side. Our core desires are either our strength, or our weakness; and all desire in mankind to accept evil, started then and there as one soul ate a forbidden fruit. The list of accepted evils by mankind has grown again and again. It has incrementally creeped up to where the greater forbidden abominable evils of sexual perversion and covetous are now desired and proclaimed to be good, and indoctrinated into children. Abominations are now accepted, and even preached in the pulpits of many false churches. Evil means, it is not what GOD originally created according to *his* desires. How can one accept preachings against Scripture? Because they seek a false covering from GOD. In their core, they hope to escape accountability by blaming those who mislead them; but this will not work. Scripture actually says, these will be punished the worst. Those who refuse to study and accept *all* Scripture, in not wanting to know the truth, commit a second greater act of pride and conceit in an attempt to fool GOD. GOD says he hates this hubris. While growing and nourishing our good and true desires, our not knowing all wrongful desires is our enemy. We must always understand and remember the place and part our free-will has in our eternal conscious reality. Love must be given of free-will, and by a true real choice; or it is not love. So free-will and choice must forever remain. Any forced appearance of love is a contrivance and coercion. FATHER cannot make anyone love him, ever. But he has done all he can, in his Holy Righteous ways, to show us *how* much he loves us. And in this diminishing time of *our* choosing, GOD now mostly looks past this fallen creation to see only the fewer bright souls striving to come to GOD. GOD senses the overall horrendous condition of the world, but they do not watch every soul's, in the moment errant and evil thought or act. That full review comes at our final judgment. There are our lower and FATHER'S higher form of omniscience. GOD'S omniscience, need not know all free-will evil errant things before, or as they happen. GOD'S "omni-science" (meaning all true science) is in their knowing every *true fact* about creation and life. All their eternal plans, reason, purpose, and the deep defining of all good according to GOD alone; and that is omniscience. FATHER said, he will not *hear* the unrepentant soul in the time they cry out to him for help, because they are contrary and ask amiss in selfishness. There are two ways of life, the Godly loving way to strengthen, perpetuate, and increase the Kingdom; or individual evil covetous ways, desiring and experiencing physical pleasure, that weaken and tear the fabric of reality. Each soul has our eternal part, in returning to GOD'S created ways, from which mankind selfishly fell away. Never believe that GOD works against, or changes, or over-rules our free-will desires; or coerces anyone to belong or to fit into any way of life. GOD will help us, GOD will give and show us love. A parent *can* love a child, more than a child loves a parent. But GOD loves us, as we love them. And as we reach up just so far, they reach down just so far. If we take a step towards GOD'S ways, GOD takes a step towards us. If I love GOD more today than yesterday, GOD loves me more today than yesterday. If I backslide away from GOD, I feel GOD pull back from me. Free-will, and our mutual loving FATHER child relationship is the order and our hope of the day. Does this sound wrong or foolish to anyone. If so, I ask, do you love GOD, if so, how much? Do you know GOD intimately? Do you love yourself, family, friends, and the little kingdom you have built, more than GOD, the creator of those things? Can we truly equate our loves? I always ask, is it ever acceptable between conscious entities, for one to love the loving spiritual or physical gift from another, more than the one who gave it? For that would be the use and abuse of love. Many who *say* they love GOD, do so in a worldly insufficient way, not obeying GOD'S truth. How can a soul *eternally* love another, if they do not agree in all ways of life. If I tell you that someone you do not know, has sent you a great gift of life, while they live on an island in the middle of the ocean. Can you tell me that you truly love this person with all you are? Or can you only say that you love the gift of life, and would like to get to know this soul who loved you and gave you this gift? And what if through earnest effort, you get to know this soul, and are not enamored, and actually disagree with their ways of life; then loving your own ways more, in experiencing their gift of life? Can you then say you love this person more than any other, solely because of the gift? We know love is the most intimate of all bonds. But temporary love means wanting to be *with* the one you love; while eternal love, means wanting to *become one with* the one you love. We see love is truly built upon our forceful attractions of common desires and needs to both please and be pleased in peace and joy, just like GOD. Now we all see how a lot of loves in *this* world, start, grow, weaken, and die. So, what are worldly temporary loves, and what is an eternal love? Fallen mankind will say no love can last for eternity, and in their lost state, this is true. FATHER has all his appliable forceful ways, that still cannot create eternal love; *but* once that true oneness of love is found, FATHER *can* eternally seal our state of love. Love is the most precious and meaningful of all free-willed things. Life without love has no shared meaning at all, let alone an eternal meaning. A love that is not sealed as eternal, serves no eternal reason or purpose. Remember always, that GOD in being eternal and perfect, they love, seek, and desire, other eternal perfect loving souls. The eternal perspective is found in eternal love and hope, and in eternal responsibility and accountability, and in eternal understanding and belief, and in GOD'S abiding created ways. I ask all captives who long to be free of hell and death to learn of GOD, and alter your core desires you now hold in for a temporary perspective of life. Turn to the eternal, never again feel life is temporary, or being of this world. Do this and you will begin to see and know all things through the understanding of GOD. It is the way for us to adopt the new, as we put away all our prior understanding, acquired in being of a fallen dying mankind. You will not see physical death, or the death of separation from GOD in the same way ever again. I know I exist, and I will forever. Spiritual death means separation from GOD, physical death means separation from a body. I know there is a door we all walk through in leaving *this* body behind; and in the twinkling of an eye, beyond time, be in a new eternal Angelic Body. If we are not spiritually separated from GOD, the second death no longer frightens us. Fear is of the unknown, but I clearly see what comes. How can anyone fear stepping out of the horrors of this divisive dying world, and into a commonly shared perfect peace, joy, and love. If not for the need and the work of GOD to be done here, I could without hesitation or apprehension leave now.

Please, you must desire and choose to be one with, and exist through all the eternal concepts and understanding of GOD. Through the eternal perspective, death of this body does not mean loss of life, or non-existence. Spiritual death is to be separate from GOD. Our disconnection from this aging dying physical platform, be it nightly in sleep, or its death, have nothing to do with the death of being spiritually separated from GOD. CHRIST said the temporary physical perspective is for the walking dead and dying, and the eternal spiritually focused perspective is for the living. Luke 9:57-62, "*AND IT CAME TO PASS, THAT, AS THEY WENT IN THE WAY, A CERTAIN MAN SAID UNTO HIM, LORD, I WILL FOLLOW THEE WHITHERSOEVER THOU GOEST. AND JESUS SAID UNTO HIM, FOXES HAVE HOLES, AND BIRDS OF THE AIR HAVE NESTS; BUT THE SON OF MAN HATH NOT WHERE TO LAY HIS HEAD. AND HE SAID UNTO ANOTHER, FOLLOW ME. BUT HE SAID, LORD, SUFFER ME FIRST TO GO AND BURY MY FATHER. JESUS SAID UNTO HIM, <u>LET THE DEAD BURY THEIR DEAD</u>: BUT GO THOU AND PREACH THE KINGDOM OF GOD. AND ANOTHER ALSO SAID, LORD, I WILL FOLLOW THEE; BUT LET ME FIRST GO BID THEM FAREWELL, WHICH ARE AT HOME AT MY HOUSE. AND JESUS SAID UNTO*

HIM, NO MAN, HAVING PUT HIS HAND TO THE PLOUGH, AND LOOKING BACK, IS FIT FOR THE KINGDOM OF GOD." This stark pure truth is a blessing to all who desire it. JESUS says, most souls walking this earth in their own chosen beliefs and ways, are all walking around spiritually dead and in hell, but don't know it. Spiritual death is our state of no conscious connection to GOD, and Hell is the confused state of an individual, separate from GOD and all other souls concerning the truth of reality. I advise all who are spiritually dead and in hell, to turn to GOD, and *escape* these most undesirable conscious states. Proverbs 15:21-27, "*FOLLY IS JOY TO HIM THAT IS DESTITUTE OF WISDOM: BUT A MAN OF UNDERSTANDING WALKETH UPRIGHTLY. WITHOUT COUNSEL PURPOSES ARE DISAPPOINTED: BUT IN THE MULTITUDE OF COUNSELLORS THEY ARE ESTABLISHED. A MAN HATH JOY BY THE ANSWER OF HIS MOUTH: AND A WORD SPOKEN IN DUE SEASON, HOW GOOD IS IT! THE WAY OF LIFE IS ABOVE TO THE WISE, THAT HE MAY DEPART FROM HELL BENEATH. THE LORD WILL DESTROY THE HOUSE OF THE PROUD: BUT HE WILL ESTABLISH THE BORDER OF THE WIDOW. THE THOUGHTS OF THE WICKED ARE AN ABOMINATION*

TO THE LORD: BUT THE WORDS OF THE PURE ARE PLEASANT WORDS. HE THAT IS GREEDY OF GAIN TROUBLETH HIS OWN HOUSE; BUT HE THAT HATETH GIFTS SHALL LIVE." Yet tragically, the world in mass confusion, choose, believe, and revel, in their ways of embracing death and hell. They speak and sing boastfully about not fearing, but of enjoying hell and death together. They employ decorations of skulls and symbols and songs of evil, death, and going to hell on highways. Are such confused thrills of this world of any value? No. They are forces of illusion that lure souls into self-delusion. People love a good magic show, but why? Is there thrill and comfort in being profoundly deceived? But when the trick is revealed, they are left asking *why* they were fooled. The true reason *why*, cannot be known unless we want to know. Can you ask yourself, why would you want to be deceived, and enjoy such a thing? The reason is that most souls do not want to know the final answers to life, which all lead to GOD or dcath, two things most do not want. They need and cling to an uncertain mind-set, denying the forceful conscious rules of reality. Most need to be deceived, because down in their core they hate the spiritual truth of it all. In remaining confused as to the truth of reality, they feel more unaccountable. We must raise our objection, and reject every attraction to deception. Once we are under the protections of GOD, Satan can no longer try us without permission. I grow closer to GOD, yet sometimes I sense a little slip backwards; so, I know I am not sealed yet. Why has GOD allowed *me* to approach him, for in my youth, I was foolishly preoccupied with this world. But at the age of twenty I heard a messenger and I reached out to GOD; and GOD was still there in a most profound way; but why? I never reached out until I was so confused about GOD'S love, and felt I had stepped too far and lost GOD'S love. I felt I had been far too unfaithful and irresponsible for far too long. But in my shameful prayer and repentant turning back, as fast as my hand reached up, GOD grabbed it tight. I get such a shameful feeling if I stumble, and must reach back up; yet to date, GOD has always grabbed my hand. But I cannot know why, because I can't qualify or quantify my growing love for GOD against my dying love for this world. I cannot measure these, one against the other. I know that if my love for anyone else, rises close to or above my love for GOD, that can only come by losing my hold on the eternal perspective. And in doing so, my hand in GOD'S must have let go; so, GOD'S hold on me must also release. The only evidence I have of GOD'S love for me, and my greatest love being for them, is in my living by their ways in their eternal perspective, and receiving their revelations through that perspective to carry-out GOD'S work and will. I have said in other books that love does not endure by its great dimensional measures, but only by its strength and purity. It matters not that I am so small, or GOD is so vast and great; pure eternal love is Godly love. Just as our love for our newborn child is purer and stronger than for a large person. The forces of evil are at work among all mankind and the angels, but never doubt the far greater immutable forces of GOD are at work towards their eternal purpose. Be not dismayed, for if we are immutably one with GOD, they are immutably one with us. Do not be partial, do not keep one foot in the dying perspective and ways and of this world. Do not think that having only one foot in GOD'S ways will bring or grant anything eternal; it won't. Being divided or lukewarm in desires for wanting and trying to have both worlds, brings only more judgment. FATHER will see and hear all, but here in our moment of choice, he mostly turns his eyes and ears away. A day of fear and dread for all is coming. FATHER *created* what is true and right; and only he and his children can know and be true and right. Any who disagree with any single thing, or in any single way, with what GOD says is true and good, is partial, inadequate, and destructive. Will I or you be among the lost; I hope we are not. Let us go where we are spiritually called and created to go. Let us do what we are spiritually called and created to do. Let us be what we are spiritually called and created to be. And let us all live and love as we are spiritually called and created to live and love. The seal of protection and election we desire, comes upon those who come *all the way* back to GOD, and comes in our differing moments. To some it is attained while still in this world, to others it comes under the throne of GOD where the Martyred await, and others during the Millennial Reign, or before GOD'S Throne. It comes at our reaching a conscious *loving* point, and not by any endured moment. I have studied how and why it is that spiritual love, hope, and faith, are the supreme forces. And how and why physical hate, fear, and pride, are love's antithetical forces. Yes, all force is born of desire, and a law of reality among free-will souls is they spread their beliefs; but projecting eternal love is our only way home. The endless contrary lustful desires of hate, pride, and envy, spread their dying confusion to others, as others spread theirs back. As these souls reach their final state, they all battle and destroy each other forever. These souls do not value or hold sacred the life within them. From our eternal perspective, we see their contrary forces can't bind anyone together; they only add destruction to all they see. But once all else is destroyed, they fix their destructive desire upon themselves forever. In their great efforts to try and hold onto anything or anyone, these souls alter their lies, again and again forever. There is only one forceful mechanism that aligns and anchors the understanding and love of many in one eternal purpose, belief, and way of life. Yet weak believers in GOD, still have faith that mankind can improve and rule themselves. Even as we see hate, pride, envy, and divisive insanity grow and rule over every individual soul in this world. Individuals are in a constant mutation of belief and understanding without any true foundation or direction. This leads to insanity, and angry denials of being insane. Can anyone who is sane, believe that billions of souls who are constantly consciously mutating differently, can ever find the truth, or live-in peace. We have to look at this continually disintegrating world of death, and its increasing societal dissolutions and ask where is any slight evidence that things could ever get spiritually better. Don't throw out abstract banal words like optimism, as peaceful shared love can never exist under any one-world government of coercion. But ever so oddly, many believe that if the world can just get rid of GOD'S truth, ways, commandments, and love, they say are bigoted, the world can find its inclusive peace among all who forever disagree more and more. Without GOD'S aligning ways, the world will find only more anarchy and ever-growing immoral permissiveness that leads to cities ruled by the same insane zombie sexual animalistic drug addled brutality that reigned in the cities of Sodom and Gomorrah. Cities where the sadistic, physically strong, and sexually perverse ruled, and the innocent, sweet, young, and weak, are abused, brutalized, sodomized, and devoured to satisfy the vile appetite of those rulers. This world's current condition cannot be reversed, the majority can never stand up and say let us return to a prior state. Only the few who follow the guidance of GOD will return to the true and good. Look at every brutal ruling power on earth, do they *ever* bow to the will of those who helped them gain power? No. They just instill divisions of fear among them to keep them disunified. We see the weak minded, always want a brutal force to take power from those they disagree with; to vanquish those that trouble their soul. They think power will then be given to them, but it will not. They never realize that they and their loved ones are pawns and a means to an evil liberal end, and they are destined to be trampled upon. The first thing evil liberal forces do when taking power, is turn to let all know their place of subservience. Now currently, evil does appear on the surface to be the stronger force; but that is only by not seeing the true restrained force of GOD. If all mankind, in all their conscious strengths and reach of force, came out at the same hour to stare at the sun, willing it to change it slightly, it remains the same. But if FATHER looks at the sun and effortlessly thinks to change it, it changes, as would the entire universe. Do any think that all evil souls in the entire universe combined, can accomplish one tiny eternal thing *against* the person and force of FATHER. Any soul that believes so, is a soul in which the forces of envy, pride, hate, and ignorance, reign. Many souls in history have known the things reiterated in our books. But most have refused to let GOD put them all together in their correct way. In the lives of the rebellious, the eternal truth of life is rarely considered, let alone lived. Now Newtonian Physics, says the mysterious force of a greater mass, attracts all smaller mass to it. But Einstein's Relativity says that greater mass warps time and space, so this warping by large objects, causes smaller objects to fall towards the center of that warped space. But in truth, everything external to our souls is just consciously and forcefully *entangled* at its creation to exists in its ways under the forceful conscious intentions of FATHER, and to move in allowance of how and where it all was intended. This allowing free-will souls to be entangled to his creation in limited free-will moving ways. As a theoretical mathematical formula, we replace the force of a larger mass, or a greater warped-space, with FATHER'S desired projected pattern of creative forceful entangled intent; *replacing* both physical force causes, with his forced conscious cause. A third way to see it, after Einstein saw a second way, after Newton saw it his way. Conscious force entanglement of all reality is easily seen by any who does not close one eye, and squint to blur the other, and crane one's neck in an effort to look around it, preferring an alternative. And I assure you, your desired theory can make all sense to you. But a physical perspective, to a relative degree, will be ambiguous and scientifically problematic in many areas, and you need a physical miracle to get started. But if you truly desire it, you can accept its ambiguity, problems, paradoxes, and miracles. Free-will desire can accept *any* belief in a soul gradually going insane. Look at a wildly insane soul, flailing and railing about in the street; they started their journey on one big false belief. Any accepted measure of insanity only leads to more and more of the same. And those who go down their own path of individual insanity, rarely realize they do so; but the few can recognize this before it's too late. So, just as our patterns of projected conscious force, create and entangle our endless quantum perceptions within our soul's sphere. The universe was originally projected, entangled, and perceived by FATHER in his great hope that it would exist as he intended. Why are we held to the ground in our created sleep dream perceptions; is it gravity, warped space, or is it our entangled particle-wave-based dream perceptions. And if we desire to fly in our dreams, how do we negate gravity, except by the same. Our desires in our dreams are projected and entangled to be what and where we desire. The manner of every *particle-wave* perception created by our souls, can be traced back to our forceful desire to exist, projected into the aether in the patterned way we want each perception to consist. All our perceptions and FATHER'S desired firmer creations, are projected conscious creations. Our experiential existence is created in our internal conscious struggle over the ways in which we desire to exist. So, FATHER'S desired external existence is under assault, by the forceful contrary ways that individual souls want to enjoy his firmer creations. We indwell our perceptions at our most intimate level; while secondarily indwelling and affecting the firmer perception of FATHER, at a shared level. We see entropy and dissolution and loss of designed complexity, are the effect of everyone's forceful projected disorder, altering the initial created dynamic ways. At the beginning of FATHER'S perceived-external reality, his set unchanging level of projected desire, sustained his perfect *dynamic* system. But now that system is rejected by the many in all the contrary ways they desire to exist, so, dissolution is inevitable. Does this mean FATHER is over-powered? Of course not. FATHER allows all free-willed self-destruction to play out, as he searches the ruins for *his* children. FATHER gave us all free-will to find his love, and warned if we sever ourselves from his created loving ways, our substantive provision must be severed also. GOD must allow all desires towards hell and death to arise within all who choose them; being all who wish to stand in GOD'S place and define a different way of life. FATHER has given us our eternal soul of sovereign free-will choice, and out of our desires we must and do make our choices. But we are all accountable to forever dwell in our final sealed choice. In GOD'S warning us of *death*, being our separation from GOD'S substantive provision, it is clear that mankind is not consciously forceful or connected enough, to sustain a physical body, or their required environment. And none can dictate to FATHER that he bow to mankind's desires, and sustain reality in their desired ways. One should surely see evil can never win, in that FATHER said he would allow their self-inflicted hell, death, and destruction. And at the last, he will suddenly hasten the end of his fallen creation. He also foretold of his next perceived perfection of creation for all souls he saves along the way, being all who truly return to never turn again. The first physical creation was once perfect, yet fell so far from GOD'S desires. But do not try to imply the next perfect creation must falter like the first. That which destroys this creation, was not in its physical design. The imperfection that destroys it, comes by all the desires of the rebellious. Once he finds all who hate rebellion and desire his perfect eternal ways, FATHER has a way to seal all true found love against all imperfections for eternity. For their searched for and found treasure, FATHER will provide an ascending *spiritual* existence our minds and hearts could never imagine. One thing I try not to think about, is what will Heaven be like. I don't think I, or any should dwell on this. This sense is strong in me, as I have been taught a lesson in that whenever GOD has heard my prayer over a certain thing in life, and I foolishly try to imagine every way it might come about; each time it has come in a way I never thought of. I take this as an anecdotal loving playful way, FATHER lets me know, to my amusement, that I can *never* know when or how he will answer. It sounds silly, but in such moments, I, being an old man here, feel like a young child being playfully outsmarted by FATHER. Do you remember the amusement and amazement of being outsmarted by your dad, bringing a childhood sense of peace and safety in knowing that someone far greater than yourself is protecting and watching over you. And in being a child of GOD, we should feel so now. And to all souls, no matter your age, who have sensed this childlike experience, I am glad for you. Now all souls are on a path or trajectory going somewhere; but do all truly see, and consider where their current path leads to in the eternal sense. I see more and more through the eternal perspective, including my path. I try to see all things in the eternal way GOD commands us to. It is often a lonelier existence here in this evil world; but having an intimate relationship with GOD, unimaginably outweighs all I left behind. Most often, our worldly family cannot understand why we live for GOD more than they do. We wish every soul would return to the true eternal view of life, Adam and Eve had before their fall. It gives life greater meaning, understanding, and purpose, in our every eventful moment. Every choice takes on the implication of protecting and perpetuating eternal life and love; our every choice has a sensed eternal consequence. At rare times, my understanding of reality has felt so vast, it seems I could (in my mind) stretch my arms out and almost touch the far edges of the universe. If you *are truly* in GOD, you feel like you truly are; yet in most-humble-childlike ways. Everything is relative when spiritually understood under GOD'S greatness, as we are entangled to the entire purpose of creation and its creator. And yet, at the same time, I can feel smaller than ever, in knowing more and morc, thc grcatncss of GOD. I hope to be sealed in the eternal, and I know eternity for many is already found. I also know, that in our soul being sealed to GOD, we will always exist on the watching waiting side of eternity, as eternity remains inside of us. Eternity's full true definition, can never come close enough to touch, or pass us by. Eternity is not like time in any way; time is for all temporary things that begin and end. This may make no sense to many, as it's quite hard to describe an eternal sense. We sense an eternally expanding moment, and hear an immutable promise that in going towards our eternal horizon, there are only and always, the gifts of a loving GOD. Yet for now, we must then turn our eyes in painful sorrow, back to this world filled with fading to emptied out souls, seeking and running nowhere, filled with death, hell, and destruction. I ask GOD why souls do this, as I no longer sense or understand why. I cannot see how any ungodly purpose, reason, or belief, can exist within any definition of sane logic and loving compassion towards GOD and one another. And I won't ask any fallen soul if they can give a sane reasoned purpose, for living such a temporary way of life, because I know not one can. Free-will, being our core chosen desires, when first firmly formed through selfish considerations, usually bring the saddest course of events. I know that this too is an experiential-qualia; and in moving from it, I hope I never come to chase such false things again. I do not claim I have *never* experienced pride or envy in some measures, but I say those were never my strong core desires. They were weaker foolish surface desires in my youth, wanting to know and be known in a world where I did not belong. These are recognizable evil qualities and quantities, I still in lesser ways, must work to completely deny. I do not find myself impervious to *all* such things, I am yet set among the vulnerable and susceptible, and those not yet sealed. But I know GOD has commanded we all strive to be perfect, so, I know it *is* possible to build and achieve that first principle. We are all in a battle for our eternal lives, and must battle against the worldly remnants within us. We must never turn back, we must move on from the first basic steps we take towards GOD, on to every next perfecting step to come. And always desire, seek, and find our greater understanding. Heb 6:1-20, "*THEREFORE LEAVING THE PRINCIPLES OF THE DOCTRINE OF CHRIST, LET US GO ON UNTO PERFECTION; NOT LAYING AGAIN THE FOUNDATION OF REPENTANCE FROM DEAD WORKS, AND OF FAITH TOWARD GOD, OF THE DOCTRINE OF BAPTISMS, AND OF LAYING ON OF HANDS, AND OF RESURRECTION OF THE DEAD, AND OF ETERNAL JUDGMENT. AND THIS WILL WE DO, IF GOD PERMIT. FOR IT IS IMPOSSIBLE FOR THOSE WHO WERE ONCE ENLIGHTENED, AND HAVE TASTED OF THE HEAVENLY GIFT, AND WERE MADE PARTAKERS OF THE HOLY GHOST, AND HAVE TASTED THE GOOD WORD OF GOD, AND THE POWERS OF THE WORLD TO COME, IF THEY SHALL FALL AWAY, TO RENEW THEM AGAIN UNTO REPENTANCE; SEEING THEY CRUCIFY TO THEMSELVES THE SON OF GOD AFRESH, AND PUT HIM TO AN OPEN SHAME. FOR THE EARTH WHICH DRINKETH IN THE RAIN THAT COMETH OFT UPON IT, AND BRINGETH FORTH HERBS MEET FOR THEM BY WHOM IT IS DRESSED, RECEIVETH BLESSING FROM GOD: BUT THAT WHICH BEARETH THORNS AND BRIERS IS REJECTED, AND IS NIGH UNTO CURSING;*

WHOSE END IS TO BE BURNED. WE ARE ENCOURAGED TO IMITATE THOSE WHO HAVE INHERITED THE PROMISES OF GOD. BUT, BELOVED, WE ARE PERSUADED BETTER THINGS OF YOU, AND THINGS THAT ACCOMPANY SALVATION, THOUGH WE THUS SPEAK. FOR GOD IS NOT UNRIGHTEOUS TO FORGET YOUR WORK AND LABOUR OF LOVE, WHICH YE HAVE SHEWED TOWARD HIS NAME, IN THAT YE HAVE MINISTERED TO THE SAINTS, AND DO MINISTER. AND WE DESIRE THAT EVERY ONE OF YOU DO SHEW THE SAME DILIGENCE TO THE FULL ASSURANCE OF HOPE UNTO THE END: THAT YE BE NOT SLOTHFUL, BUT FOLLOWERS OF THEM WHO THROUGH FAITH AND PATIENCE INHERIT THE PROMISES. FOR WHEN GOD MADE PROMISE TO ABRAHAM, BECAUSE HE COULD SWEAR BY NO GREATER, HE SWARE BY HIMSELF, SAYING, SURELY BLESSING I WILL BLESS THEE, AND MULTIPLYING I WILL MULTIPLY THEE. AND SO, AFTER HE HAD PATIENTLY ENDURED, HE OBTAINED THE PROMISE. FOR MEN VERILY SWEAR BY THE GREATER: AND AN OATH FOR CONFIRMATION IS TO THEM AN END OF ALL STRIFE. WHEREIN GOD, WILLING MORE

ABUNDANTLY TO SHEW UNTO THE HEIRS OF PROMISE THE IMMUTABILITY OF HIS COUNSEL, CONFIRMED IT BY AN OATH: THAT BY TWO IMMUTABLE THINGS, IN WHICH IT WAS IMPOSSIBLE FOR GOD TO LIE, WE MIGHT HAVE A STRONG CONSOLATION, WHO HAVE FLED FOR REFUGE TO LAY HOLD UPON THE HOPE SET BEFORE US: WHICH HOPE WE HAVE AS AN ANCHOR OF THE SOUL, BOTH SURE AND STEDFAST, AND WHICH ENTERETH INTO THAT WITHIN THE VEIL; WHITHER THE FORERUNNER IS FOR US ENTERED, EVEN JESUS, MADE AN HIGH PRIEST FOR EVER AFTER THE ORDER OF MELCHISEDEC." Such souls, who hold and endure to the end, receive GOD'S loving rewards. It is our battle, and all battles are fought on battle fields. This term battlefield speaks of the physical dimensions alone, but the battle itself is more spiritual, emotional, and psychological. We know this world was once not a battlefield, as in the beginning there was no battle here. In this Creation's beginning, FATHER perceived all things to exist according to his omniscient eternal love, and every perfect desire. And each free-will soul, was gifted ready and capable to learn and accept the ways and things of GOD in love, or reject them in hate. All Creations of Heaven and Earth stood for a profound moment in the initial spiritual and physical perfection of nurtured certainty on the part of FATHER; but in that he was seeking to find true eternal free-willed love and belief, the possibility or probability, born of free-will and chance must be real and set as one disallowed choice within that same sustained moment of perfection. At one moment, in the heart of a certain angel named Lucifer, that *very* first iniquity was born. Lucifer then fell and fell in his soul, and *spread* his weakness of confused pride and envy to one third of all the angels in heaven. Lucifer also knew the gifted free-will soul called Adam, had dominion over all the earth, for GOD to watch him learn what he would love and choose to do with it. And Lucifer desired the earth, wanting Adam to relinquish that dominion to him, and for Adam to fall with him into spiritual hell and death. GOD'S creation of the Angel's ethereal bodies of life in heaven, and the firmer physical bodies of life in the earth, were created in hope that the eternal ways of a common spiritual love would be desired and lived by all within both realms of differing physical experience. We now expect a blend of both, in one final new realm. This has been no small effort by FATHER. Scripture says FATHER *does* get wearied by such great rejection in his search for love. Scripture says GOD works, and that he rests on the *Seventh* day. Can FATHER get physically exhausted? No. But he desires and needs his conscious peaceful love and rest from the mounting conflict and wasted tedious work. The only way GOD is wearied, is in being rejected and abused, bringing hurt and tedium by the rebellion they encounter in their pursuit to find an eternal family. We are taught how the initial universe was designed, layered, and created during earth's first six rotations. And how in FATHER'S desired reasoned purpose, it was contemplated, planned, and finely tuned to be dynamically complex and beautiful, before his step-by-step creation of it all. And upon its completion, the mystery, wonder, and uncertainty, of Adam's blank free-willed soul was set loose within it; as this was its purpose. In Scripture, we see GOD so hopes for what the Angels and Mankind will choose; wondering if or when a possibility or probability of being rejected will rise to a level of free-willed certainty. And FATHER clearly says he is wearied by what has freely occurred. Now Scripture does not say how long Adam, and then Eve lived in GOD'S Garden before falling and introducing the disordered temporary aspects of time into a once perfect place GOD had prepared. We see mankind's individual evil desire, first arose in the weakness of Eve; and by Adam's love, her beckon held a strong influence over him.

And at the moment Adam's unfaithfulness was known to FATHER, the Earth lost its eternal purpose. And Adam and Eve's body and the whole firmer physical creation lost its purpose and began its fall of stretching beyond FATHER'S set perfect care. Falling into a downward devolving spiral towards ever-increasing pain, sorrow, insanity, hell, death, and dissolution. FATHER, who literally held the universe in its dynamic perfection between his hands recoiled, pulling his hands back and away at hearing this painful rejection of his love. And the physical universe was stretched from its initial dynamic perfection. We understand by FATHER'S omniscience that his creation was formed in his love to contrast all hate, and in his light to contrast all darkness. Love and Light are firm Scriptural foundations FATHER chose to build his good intentions upon, and to contrast all contrary evil intentions of others towards his desire. But we now see mankind's terms of black, white, and growing shades of *grey* are used, trying to allow a non-existent blurring spectrum of the division between good and evil across the middle, as most now refuse to see all things as good or evil. Mankind sees the illusion of an expanding grey spectrum called individuality, that now dwarfs the pure good and pure evil. They see a false spectrum of life, where true knowing and accountability do not exist. But we see rebellion, being individually defined in endless immoral shades, as the confusion of individualism grows and spreads. Remember, in *his* perfection, JESUS sees all things as being the brightness and direction of eternal light and life, or as a void of devolving darkness hell and death. JESUS sees no grey area; he sees all are accountable, and at the last he will see all souls in their free-willed freely-chosen state of good or evil. GOD does not divide life, existence, or reality, all division comes of itself. We must choose to grow our eternal responsible faithful love for GOD, or a rebellious individual hate for GOD; again, there is no grey in GOD'S Judgment. There is no allowed purpose or reason for sin or rebellion, and no individually defined morality. At the last when all good and evil souls have become clearly and irrevocably divided by their desires and choices, and a true eternal freely chosen chasm opens and appears between all souls. Then the eternal-divide FATHER will rule over, has appeared of its own means. FATHER then states this obvious outcome in his final decree of Judgment, and partitions his final eternal physical/conscious found and lost realms of reality. So, let us seek the beauty of our eternal freely chosen oneness with GOD, let us desire, know, and experience the peace, joy, and love, of this oneness. Let us want and work to acquire and build this oneness of hope within ourselves by GOD'S loving gifts of promise. Or be an individual, sinking into an increasing loneliness of despair and anguish, void of every tiny physical comfort and spiritual joy. Being forever filled with unfeeling, unresponsive, selfish wants, that are never met, rambling desires never fulfilled, and hopes that never materialize. Do you or I, unknowingly deep within our core, love our selves or people or things of this world, more than we love GOD. I tell you a most firm immutable truth, that only GOD can truly judge this in us. We must strive and grow in our faith and hope that in his day, GOD welcomes us. Then and only then, will we truly know. GOD says, we are not a *judge* of declaring the final outcome of ours, or anyone's soul. Yet many souls interpret this command, to say we are not to judge what is sin, or a sinner. That is ridiculous; we are commanded to judge all sin under HOLYSPIRIT'S interpretation. Play a rebellious god, in ruling over your life for this moment, if this you desire, but ready yourself for what forever comes by it. We must ask ourselves, are we pure, bright, and loving to GOD, or partly, or mostly dark and hateful to GOD? Do we see ourselves as grey? If so, what tint of grey do you see yourself to be? Can you truly see yourself correctly? Does GOD see you in the same tint as you do? Do our senses bias towards self-preservation? Do we repent and remove all our darkness as we find it? Are we growing brighter, perfecting our ways towards becoming one in GOD'S ways? The darkness of hate consumes all creation beyond the Holy of Holies, and the measurements of dark energy and dark matter now comprise over 94% of the observable creation. What do you think this form of dark reality is? Astrophysicist says they detect this dark reality through gravitational lensing and universal inflation; yet have no clue as to what it is. They also cannot measure the universe accurately to see this darkness is both growing and accelerating. This universe has a little over 1000 years left (to include CHRIST'S Millennial Reign). And if you think the world is gone insane now, just wait. No soul since Sodom and Gomorrah has seen anything like the coming insanity and depravity against GOD'S ways, that will be set upon the young and weak. Remember, only those with GOD will see this for what it is. Even the mildly insane, will only see a growing godless physical normalcy. Their increasing inner darkness, brings them the appearances of time, and their approaching end of it. Can anything drive people insane and enrage them more than realizing they are eternally lost? So, how dark and dying will it all become before the end? I can only say beyond doubt, that the numbers of people who die of stress, fear, suicide, drugs, murder, famine, disease, war, and the cleansing of GOD, will over a short twelve or more-year period, kill most all human life on earth. The last three Trumpets of the Tribulation, are as was with Sodom and Gomorrah, a cleansing of GOD'S worst enemies in mankind by GOD'S Holy Angels. This, in preparation for CHRIST'S Millennial Reign upon a dying earth. Again, this Millennium is the final dispensation for trying certain souls. It is the last chance for all souls, not yet found to have made their final choice. In these current last few decades, earth's population might creep up to nine or more billion. But it will be reduced to maybe three or four percent of that, as we enter CHRIST'S Millennium, but we are given no specific number. We read that first one fourth of those alive die, then one third of all those remaining die, and then many more die. Then one third more of those still remaining die, and then many more die. And then Scripture says these days of death will be shortened by GOD, or no flesh would be spared. Then after CHRIST'S thousand-year rule, Scripture says the earth in measure, has been repopulated, and again, most reject the perfect ways of GOD and gather from the whole face of the earth to battle against CHRIST, and FATHER destroys them all with fire.

Revelation 20:9, "*AND THEY WENT UP ON THE BREADTH OF THE EARTH, AND COMPASSED THE CAMP OF THE SAINTS ABOUT, AND THE BELOVED CITY: AND FIRE CAME DOWN FROM GOD OUT OF HEAVEN, AND DEVOURED THEM.*" This all speaks to the standard and quality of eternal love towards them that GOD seeks. A loving desire for GOD, most just can't hold, and so cannot freely give. At their core, these just cannot find GOD to be their greatest eternal desire. Some come to know each, and have an equal eternal love for the Trinity of GOD. Some love GOD in a fairly substantial way, but their love for GOD is just less than their love for certain people and the things FATHER has gifted and created us all upon the earth, but the greater portion of soul's core desires are for the bodies and things below, besides, and around them. And GOD'S great love from so high above, is just not enough for them. They just cannot be that, as it is just not who they most desire to be. GOD has sorrowfully found that among mankind, it is not the many, but only the fewer that come to love him in his eternal way. If I fail at the last to be the loving standard of GOD, can, or will *I* see unfairness? How can I, in my being the soul who has made every choice for myself, turn and blame GOD. One can in anger and envy blame GOD, but that can never make it true. All fault will assuredly rest in our core desires and chosen intentions. And those who desire and see GOD'S full goodness, and put forth their responsible faithful efforts and love for GOD to find the great beauty of perfect love and peace they crave; it is just who these souls freely are. By their own strong loving desires, these just are the true Children of GOD. Will I or you be there and rest in GOD'S perfect judgment? I only know that GOD'S final word will be pure, true, perfect, and eternally enforced, as it is the immutable promise of GOD. Those who find his reward, are all who in great love *and* fear, fervently watched, spiritually prepared, and steeled themselves in love. Our eternal choice for GOD is not something that can be approached casually or irreverently, or be a secondary focus in life. This must be our core concern in life, it must be who we eternally are. Now we read the conscious projected forces at play in the last ten or so years of this last dispensation of choice, will be projected on a shocking scale, not seen since FATHER'S creation of earth and all life itself. But this approaching battle of great forceful intent, is coming from both sides in contrary waring good and evil ways. Now FATHER knows and holds a true emotional sense of the *sum* of abusive uncaring conscious projection against CHRIST'S great suffering sacrifice. This stokes FATHER'S greatest righteous anger. Such souls are set for a day of FATHER'S wrath. We see before FATHER and his wrath come, JESUS has been so abused by mankind that FATHER is pushed beyond the edge of his compassion. And what comes upon such rebellious evil souls, is their *just* eternal banishment, in their long over-due day of judgment. I have my hopeful loving understanding of the eternal perfection of GOD, and my great fear of losing this same love forever. I am shamefully aware of this world, and weary of it; but we must be vigilant in knowing and controlling the forces we project. Evil most often comes vaguely and subtly, or profoundly, seeking out any weakness to exploit, to set a faint evil root to grow. There are the evil forces and chance occurrences we strongly see before and around us, and other devised evil forces being a little distance off and coming our way. There are those evil intentional forces that are yet far off, that come fast, seeking to pull and push us downward. We exist at the shoreline of wanting and choosing the good, while evil wave after wave crash against us. In such an evil world and day as this, evil intention and chance occurrence surround us from every way and means. Yet we are to follow and be one with our LORD, standing in peace and joy at the center of GOD'S love, it being an impenetrable ring of fire.

And so, our eternal perspective and our diligent discernment and awareness of it *all,* is always called upon; which only comes by the indwelling HOLYSPIRIT. Remember that he who is within us, is greater than all others in heaven and earth combined. And if we are in GOD through all terrors and troubles, we have their peace in our soul. I assure you that FATHER is not worried about what comes, he does not need to consider such irrelevant things; but he does hope for us, to in his image, stand strong against what comes. FATHER watch his SON stand firm against unimaginable evil and suffering. FATHER wants us to overcome weakness through him in our striving, hoping, praying, adopting, and allowing the great forceful eternal perspective of GOD to lead us in all ways. Surely any honest soul sees that to become one with and in GOD in reality, must entail our being one with GOD in all existential ways. And this oneness makes *us* stronger and happier than all other things known or experienced combined. All the unknowing souls, practicing partial weak ways of believing in and loving GOD, will hear and say *we* are fanatical, taking things too far. Please see you can *never* go too far towards GOD; but know beyond all doubt that you *can* go, *not* far enough. How do we reveal to this world that it is our desires to exist, that are our projected forces of life and creation. I ask, have you ever truly moved, thought of, received, or gotten rid of anything, which was not at first motivated by your desire of intent? Can you move, or think of, or cause anything, without first forcefully desiring and intending to do so? No, you cannot. Can or have you ever desired anyone, or intended anything in a non-forceful way? No, you have not. You cannot even perceive yourself to exist, with*out* your projected forceful intentional desire to do so. Not the first or tiniest physical aspect that exists, has ever moved, without a soul's projected force of desire initiating that action upon it. Know this truth, or believe any lie you desire, but you will exist eternally through either your good fulfilled, or evil denied desires. You will always forcefully desire, as it is who and what we as souls inherently are and do. I find GOD'S truth and love are so powerful, amazing, and immeasurably more assuring and peaceful than all the worldly loves I have ever felt combined. I am so strongly drawn into it, and held so strongly by it. I feel I could never be dissuaded or separated from it. Yet I fear and must guard against any such confusion in my every remaining weakness. Does this sound like apprehension, or unsurety in me? It is not. It is my strong knowing realistic desire to protect that which I love. This is the basic nature of all and every fear.

Fear is our desire to not lose something we love. The stronger our love, the greater our fear of losing it. Be it for GOD, or our life, or the souls and things we value within our life. Our levels of our love and the fear of losing it, are always one and eternally inseparable. Do you want to know how much you love GOD or someone or something? How much do you fear losing them? There is your answer. Most fear losing their physical life above all things, which is the seat of evil. Are temporary aspects worthy of such fear? How much do you fear losing GOD? Our qualitative and quantitative equation of love and fear reveal whether our desires are of a temporary physical perspective, or an eternal spiritual perspective. If a gun was to the head of the one you love most in *this* world, and you were forced to choose between losing them, or GOD, who would *you* choose? A long introspective contemplation, and a great honesty would be required to answer this. It is our force of desire that pushes the brake, or accelerator, and also holds the wheel that steers us around in this world. It is our soul who receives physical information through the sensing body/brain, and our soul's projected forceful desires that create our perceptions and instigate this same body to move. But the lively heart-beat, breathing, and so many other things required for the body to live, are not by our forceful projected desires. These come by the funneled down complex forceful patterned desires of GOD. Every time your heart beats, every time you open our eyes and see, every time you feel, touch, and know reality, you must be thankful to GOD for providing and sustaining your sensing platform of existence. Now we all once were, or still are, by the balance of our loving desires, unworthy and rebellious against GOD. Do we see that by the laws of conservation, once they are projected, our intentioned forces, in their degrees, must converge or interfere to weaken or strengthen all other interacting forces to the same degree. So, our force now exists in endless projected empirical and spiritual ways. Do we see our force that exists in our words, being our force of meaning and understanding and belief, that move us and affects others. The forces of love that stir and change and draw us. Do we see how our projected forces interject with the currents and movement of *all* that exists around us. Every spoken, read, or heard word, forcefully affects in at least some small way, every soul it touches. We all sense nothing more than forceful pushing and pulling in different spiritual and physical ways and directions. We see forceful words push and pull to align conscious coalitions. We see true and false understandings pushing and pulling as souls build true and false beliefs and alliances accordingly.

All desires of love and hate, *are* projected into the physical and conscious sides of reality, and the following associated conscious trust and mistrust *are* secondary projected forces. When souls speak to me, I always sense whether their forces are positive or negative towards GOD. There exists no gifted soul or physical element, which at its first existence, was not brought by the spoken purpose of GOD. But it has all now been greatly hijacked by rebellion to be used, abused, and remolded into images of individual purpose. We think by force, we speak in force; all information rests and is shared within all projected patterns of force. All simplicities and complexities are aligned by patterned force, and all old or young force was first newly generated and projected by GOD'S or our desirous intent. Our projected forces act upon the most basic smallest singularities of the aether to build our complex creation of a perception, points not detected by mankind; and by this fact, we see no soul can hold a provable understanding of *any* single physical thing, minus a provable understanding of *every* single physical thing. We must accept that there has always been and always will be, the initial eternal conscious creative ways of reality and life. And there is nothing more besides, before, or above this understanding to be known. Resist the secular confusion of what physically existed before that, and before that... No truth of reality is found down that path, only endless illogical meanderings into the realm of delusion and insanity. GOD is eternal, and to understand the endlessness of GOD, Scripture indicates GOD is not linear in any way. FATHER is a loving sphere of created reality and existence. FATHER is the beginning of everything, and he must also sustain it, or bring the end of it. And our soul's eternal reality and perspective are not linear, they are spherical in size and scope; and our essence and knowing awareness fills and stays and works within us. But a temporary perspective sees time and reality as having an impossible and undefinable linear form. How could reality and our existence just unexplainably start without any reason or purpose, just to soon forever end? It cannot. The wisdom of reality rests in that FATHER has always been, and we all will forever exist *within* his reality, and nothing has ever existed beyond him. We can only know that in the stream of FATHER'S eternal desires, he has divided out, to us, our eternal creative soul, and shares with us as much truth as we can know by the measure that we love and understand him. FATHER shows no restraint or reservation in sharing love and truth, which are never *first* found in us. Set no limits as to what you wish to know of GOD in love, and GOD will meet, know, and love you there in that same measure. Do not wish to be met or loved insufficiently, but completely. We each by the gift of GOD, now appear upon his eternal stage, as eternal actors without end. We can know nothing, until we fully grasp how our tiny soul was divided out and gifted, we may even say birthed out from FATHER; not as who we each became, but as blank spiritual probability equations. There and then, at our first sensed forced projection of awareness, we became a learning, knowing, loving, aware part of reality without end. Now in our measure and moment, we can never have an *experiential* understanding of any very initial aspects of this external reality, as every first aspect was before we are; and we are not a cause of any first aspects of this external reality. But if we understand enough, we know that the *cause* of reality's initial state, will also be the *cause* of its last. The beginning of any quantum creation must be entangled to its eternal existence, or the last of it. The conscious source of any perceived state of anything, must sustain it, or perceive its end. A first caused intention of desire, is an entangled consistency of purpose that allows it to endure. The creator of anything, must be the desirous forceful sustainer of it. The conscious *creator* of it, must desire to keep *remembering* it; and the giver of life, must also be remembered. Run from and forget this all if you will, but the truth remains entangled in every way, even in every soul's denial of it. None can break free, and become their own eternal truth. Embrace that you are an eternal creative soul, and examine and see clearly what you have become and created with it. Seek to know what it is you were and are meant to be in a glorious fashion. There is a perfect reason and purpose within GOD for his spiritual gifts and physical creations. FATHER'S creation is eternal against all contrary things, being external; but not against all free-willed contrary forces, attacking from within. Eternal joy, or never-ending sorrow lay before us, what do you desire? And why? Desire FATHER'S complex possibilities, and certainties of hope and life that come from him alone. I can tell all beyond doubt, that FATHER gifted us each in love, hope, and want. GOD waits and watches for us each to spiritually come back to them in love. When GOD has to leave souls behind in their eternal evil contrary desires, to sustain GOD'S own loving eternal desires; do not doubt that they do this in sadness. GOD must be GOD in perfection, so embrace this truth. No one can take from, or bear GOD'S omniscient, omnipresent, omnipotent, eternal burden. We need to see and accept what and who *we* are meant to be to GOD, as far lesser souls. It's a good loving and most peaceful thing, to find the joy GOD has prepared for us. This sounds so irresistibly true, strong, and perfect to me, and I cannot grasp it being rejected. I ask myself how *strongly* can a soul love and cling to a dying body and temporary life filled with inanimate unloving provisions, it is beyond my mind. How can a soul love and want most, that which is temporary, that they know must fade, die, and be gone. I hate the cruelty of spiritual death, so, I strive to hate all temporary seditious unholy pleasures that pull souls down. All souls who hold to a temporary perspective, live for and seek evil temporary pleasure. And holding a temporary perspective, they ask how can one hate temporary pleasure. Well, by knowing and hating what those seditious pleasures eternally bring, and knowing there are the exquisite eternal spiritual pleasures. I ask again and again, how can anyone love and choose pleasures that are illusory, hollow, and void of eternal value, over the creator of them. Such pleasures rob the soul bit by bit of spiritual strength, grace, vision, love, and truth. From the depths of a fallen temporary perspective, one can only see the next temporary pleasure, and never its underlying cost and cruelty. But in your eternal perspective and family, these desires fade as the cruelty of all temporary pleasure becomes more and more clear. One by one, these former controls of our temporary desires are replaced by our ardent knowledge of their eternal evil. After a certain number of your temporary desires are overcome in this process, you will strongly see this as a quale and qualia of the eternal perspective. Now I have yet to overcome all things, I still desire certain sweets. Every once in a while, I walk through the bakery section and glance, but as this desire is faded, I very rarely buy anything. But as an example, over several months I foolishly walked through the bakery section of a certain store on occasion, noticing they were *always* out of apple and cherry turnovers, that still tempt me. But one day, there they were, freshly fully stocked, and I was taken, and bought a few cherry turnovers and ate them for my lunch on my way home. Now there was no sin in eating them, but it was a physical strength of desire, that must reflect an equal amount of my spiritual weakness. It shows me, physical desire can still win over my resistance. Now I did not derive the level of pleasure from eating them, I used to, it was noticeably weaker. So, in this, I see progress is made, a progress that can easily be lost. Common simple food is a physical necessity, but craving the extravagant and tasty, that which gives us temporary physical pleasure is not spiritually helpful. If something extravagant is all we have, and we have need, of course we eat it. But we should not go out of our way to choose and exercise things for which we sense physical cravings. *All* temporary pleasures surround us, but they need to be felt through the eternal perspective. And in denying physical pleasures, they are replaced by spiritual pleasure and strength. This is in part with transforming our once fallen physical creature, into a strong spiritual creature. Does this sound unrewarding or unpleasant, I hope you see and desire its eternal spiritual reward of love and peace. I spend a lot of my wakeful moment contemplating the illusion of all temporary things. Maybe I hope to find a more relevant quality of the temporary perspective that reveals a stronger way for us to fight our desires. But all I ever see is that these strong temporary physical pleasures, are just what most truly desire. And even GOD will not over-rule this obstacle called free-will desire and choice. GOD has offered us everything that is eternally good, yet most choose the temporary. I wish I could find stronger reasons that reach deeper into all souls. But GOD has *already* spoken all great and reasoned words of love in Scripture, yet they are not heard by most. Are our foolish and zealous hopes, found in not having an adequate fullness of omniscience? Maybe. But we know that in the unfolding of reality among many souls, that *all* our hopes for other souls in this world, are not going to be met. I do hope for you and I and all to find our true peaceful resolve. One can only and always forcefully speak and write in the hopes of others seeing and hearing. It is the speaker and writer's great dilemma, that most souls hear, interpret, and believe things differently. So, we wish we could say all important things in every different way for everyone. But the rules of reality do not permit such luxury. One can say things in three different ways *at most*, before entering the realm of offensive rumination where you will lose the attention and interest of everyone. One hardest part, is the struggle to adequately say what we can never quite put to paper. Some understandings are prone to voice, or pen and paper. But some far greater understandings, seem to resist being put into adequate words. I tend to think these are the things that must come to each soul by situational experiential revelation from GOD. We struggle for words to convey basic understandings of GOD, and then receive far greater aspects we will never be able to share. We find in the expanse of it all, that the *vast* entangled relative ways and purpose of reality, are too much for us to hold all at once. And only harder to explain, describe, and define with mere words. Can one pour the ocean into a cup. Now in that all reality falls under the one relative entangled purpose of GOD, a true understanding of it all must rest deep within every single aspect. From one tiny fragment of any created aspect, the truth of all reality can be seen by GOD, but not fallen mankind. And among GOD'S mature children, the fullness of it can't be adequately or properly spoken, written, or shared. I shouldn't be, but I get frustrated at the limitations of what can be expressed and received by written and spoken language. The expressions and inflections of tone and the eyes and the body, or a smile or a frown, add a greater dimension, giving souls a better ability to share more than with written words alone. But even these added dimensions are not enough, nor often afforded in broad public discourse. By seeing the whole creation of GOD in all its courses and ways, there are all the aspects I understand. Yet after years, I cannot find a way to share certain realities. I lose my peace in trying, so, I know it is not meant to be. We see the language of emotional understanding and belief, are the experiential qualia of agency, within each soul. So many things can only be known through our awareness of the experience. I, like all, have my valued experiential pages and books of these things written within me that I can never seem to share. These are what I call my greater treasures and gifts from GOD. And these are a great part of *my* experiential proof of GOD. Even if an experiential-qualia from GOD is experienced by two or more souls at the same time, these qualia cannot be *fully* described between them through language and expression alone. Spoken language that can fully share all experiences, does not exist, they just can't be properly shared in that way. But what about consciously sharing the experience itself? This conscious quality does exist, whether we all can share it or not. The subsequent affects these qualia bring us, *can* in great part be shared and enjoyed among souls upon a shared comprehension, but not their deep personal meaning. All shared things of truth and love will be experienced between all in Heaven. But for now, this is found only between certain fewer souls and HOLYSPIRIT. We must accept the basic foundations and laws of all consciously creative and created reality, as these are GOD'S ways, aspects, and properties; these are who and what FATHER is and has always been. We must also accept that a multitude of conscious free-willed entities, can from within, undermine and affect the physical creation of a vast immensely more forceful agency, like a minute virus infects the body. We see that all physical acts are not caused by, but they are enabled by FATHER'S first forceful act of creating the physical side of reality, being a *required* half of everyone's existence. Yet GOD alone holds, knows, and rules the foundations of that physical half. The physical reality has always been a requirement that has been conditionally created for us. Our soul is gifted eternally, yet we experience only a providential physical connection as GOD'S eternal gift of a body, would make us higher eternal Gods unto our self. This gift, FATHER *has* given only to JESUS, and HOLYSPIRIT whose body is the aether. All below the TRINITY, are knowing forceful *halves* of lesser creative gods; experiencing our fullness of being those lesser creative gods, by GOD'S provisions of the other half. By this, we see we are not eternal *self-sufficient* Gods, yet we are all offspring. And if we do desire it enough, we may become a eternal child of FATHER. This divided gifting of many conscious forceful souls out of FATHER, must be accepted in that we can't, nor ever will fully know this equation. Might the combined souls of all the angels and all of mankind represent a tenth of one percent of FATHER'S soul. Might it be far, far less, or more. We can only know that the first two divisions, brought out from FATHER to bring JESUS and then HOLYSPIRIT, were by far the greatest divisions. Might JESUS and HOLYSPIRIT, each be more than the sum of all below them. Might the SON and HOLYSPIRIT be a ten or more percent division of FATHER. These things matter not, what matters is that FATHER and JESUS and HOLYSPIRIT have not given any great or final control over any portion of the physical half of reality to any soul below them. So, all evil souls, will find no power or weapon or advantage has been given to them that can be of use against GOD. At the very last of Prophecy, FATHER effortlessly looks down upon the masses of the rebellious gathered in confused belief, to come against his SON. And by his forceful desire alone, he simply perceives fire to come down from Heaven, which burns and dissolves and reclaims the matter in the physical half of their existence. These rebellious souls, then sleep in their non-existent state until GOD resurrects them in their new, mostly unresponsive, unfeeling, eternal body; being their prison cell. And GOD places all lost eternal souls he has Judged to be enemies of eternal peace, joy, and love, within the most separate secure prison imaginable. Forged of a most broadly defined fission, and extreme gravity, (imagine a fiery black hole) from which none can escape. The force of will by a loving GOD to find and treasure their eternal loving family will not be subverted or delayed beyond GOD'S desired conclusive search. At GOD finding the exact moment and end of his diminishing returns, and seeing no hope for one more additional loving soul remains. Then the *end* of this current Earth and Heaven, will suddenly come. And the start of a New Earth and Heaven FATHER *will* perceive, will begin; it's just that simple. And all fallen *individuals,* do not know they are racing each other to see who can reach their prison cell first, to be burdened and weighted down the most. Try and complicate this simple truth if you desire, but you will only complicate and increase your own confusions of hell. Accept the pure simple truth of GOD, let it unravel all your false complications that confuse and chain your soul, and you can be truly free indeed. There are layers of profound understanding in knowing reality, which most souls have no desire to see. If your conscious state feels trapped within the endless confusions of hell, look up, not around; there is no help from this world in finding the truth that frees a lost soul. Join us who strive to escape in returning back to the original created ways of GOD. Now some say, if the original force of the physical creation were once perfect, how could it then become disordered. Again, only if free-will choice is allowed for finding love. Perfection can be corrupted from within, and do not forget that free-will means GOD cannot control a soul's desire to be forcefully contrary. Our willful force, is always born, completely separate from GOD'S. FATHER has gifted *and* allowed the free forceful physical expression of our every desire, to see who and what we want to be and have. Every soul's free-will forceful thoughts and actions are set beyond GOD'S control, to allow us to learn, understand, believe, and project our own forceful desires of will in any way we choose. Why do some believe GOD knows and controls their acts and thoughts; this is an insane attempt to absolve one's self of responsibility and all accountability. In Scripture, GOD *has* prepared and inserted a few men among mankind for a life-long Godly purpose, who did not have a *fallen* free-will. Luke 1:15 "*FOR HE SHALL BE GREAT IN THE SIGHT OF THE LORD, AND SHALL DRINK NEITHER WINE NOR STRONG DRINK; AND HE SHALL BE FILLED WITH THE HOLY GHOST, EVEN FROM HIS MOTHER'S WOMB*." These few, were temporary created ambassadors of GOD, whose lively purpose was to reach and turn mankind from destroying themselves. I see no foul or violation by GOD against *our* free-will in that. GOD tries to call out in many ways, so when GOD sent his steadfast Holy Angels with their short, but very specific messages, was that any different? No. It was in essence, the same form of act. To find in Scripture a few inserted examples of missioned Godly influence, cannot be used to imply that GOD controls *us all* in our thoughts and desire. CHRIST *freely* accomplished every request FATHER sent him to fulfill, was he controlled by FATHER in freely doing this? No. When GOD places something to do, upon any free-will soul, it can be rejected. Predestination is a physical doctrine of deceit, that denies the existence of free-will and love. Now we see the strength and reach of force in all souls is different. FATHER'S is greatest by many magnitudes, then JESUS and HOLYSPIRIT. Then much lower, the Seven Arch Angels, and the Four Seraphim in their specific order. Then the Cherubim and lesser types of Angels in their orders. And lastly, Mankind in our created orders. Can any of us know or find on our own, where we are created to be in these forceful orders? Not without following GOD. GOD will lead us into our perfect loving peaceful place, where we find our perfect purposed joy and our eternal belonging which mankind cannot find or have. A temporary religious perspective, thinks heaven is a perfect physical place for satisfying all physical desires. No. Eternal love is found only in all being joined in a perfect spiritual way. We need to see that *eternal* happiness, rests solely in finding our eternal reason and purpose for being. We can find no *eternal* happiness outside of being the soul we were gifted to be. Just freely choosing and taking one step beyond our perfect place, is where all our freely-chosen rebellion to GOD'S creation, purpose, and Family, begin. The loving layers in all of this are impossible to explain; they must be found through experience. At first it can be hard to see how in an eternal perfect plan, all souls must have a perfect place in order, and desire their place. Can you see that if souls want to rise up or drop down out of a perfect created order (as Satan did) saying he will rise up to be like the most-high. That nothing but disorder can occur, and all eternal hope is lost. Entropy seeks the perfect order of pure disorder, where created eternal complexity can't exist. I truly care not *where* in the orders of GOD I am created to be; this just cannot matter. Do you not long for your own eternal place, which was created for you? Do you hold any apprehension and doubt about GOD'S most solemn promise? Why would any but a loving GOD, call and hope and wait for us to come to him; only love does this. We see no forcing, no lies of manipulation, no forceful over-riding of freely chosen wants and desires. If GOD were not founded in true deep sacrificial love, they would not wait for our choosing his eternal ways, they would be forced upon us, as Satan would do. We must understand and live in the created ways of GOD by our loving desire. FATHER knows he cannot force a soul to love him. He could force obedience if that were all he wanted; but that is not all that is wanted and needed. Only a giving loving soul, wants and seeks only other giving loving souls. This is the natural law of a loving conscious reality among many forceful self-aware entities. There are not many spiritual directions: there is towards, or away from GOD. There is one single shared path for all souls running to the perfection and eternal provisions of emulating GOD; but countless individual lonely paths running from the same. A spiritual adage goes, all worldly roads lead to hell. Now some say, why cannot GOD just deny the existence of certain choices and interject some greater special force and understanding upon all souls so they would prefer the ways of GOD. To deny even one hurtful choice, limits and restricts free-will and diminishes the evidence of love. Complete free-will, equals *true* free-choice. Any lesser control over our preferences, would in its measure, interject the will of GOD over ours. And it would taint our actual choice of love. If GOD did those things, he could never feel fully loved as one great eternal family, which is the purpose of his creation. The temporary perspective centers on a goal of self, and physical desires. FATHER focuses on one common shared loving goal for all souls, and has no need or desire to sustain any unloving, or lesser imperfect loving souls in a lesser form of loving obedience. Only a freely chosen faithful projected force of longing desire can ever be felt as eternal love. When temporary loves are studied, one naturally finds they are born of temporary desires that fade and change. And of course, eternal love is only born of eternal desires that bring eternal meaning and purpose to life without ever changing. Why do the fires of temporary life fade? Because temporary desires always change and fade! Without a shared eternal love, life is reduced to endless paths of fading banality. Love founds and raises all meaning, purpose, and reason, to live. One shared pure love brings us all together, one shared pure belief and understanding binds us all together. A lonely individual life slowly realizes with age, it is losing all shared purpose, meaning, and reasons to live. Eternal love is to want peace and joy that is shared without end among those of Godly desire and belief. Eternal love disappears from this world, more and more with each day, and FATHER'S search for eternal love is the only reason it still exists. The many, not seeing their eternal soul, soon find life itself is just not worth the trouble. They do not see the difference between being an ever-diminishing individual, and being part of a true eternal loving GOD. Don't let the new young heated appearance of the temporary dying loves found all around this world fool you. Eternal love is by immense measures, the sustaining force of all reality. Fallen souls form individual definition of love, and live in all the weak confused individual ways they desire and choose. The senses, emotions, and understanding of eternal love, cannot be held, without first wanting GOD'S ways above all things. Only an eternal love for GOD *and* his ways, can enter eternal life. The design is fail-safe, no one in any simple curiosity can try to sample these things; and their doubts will only be reinforced in any foolish effort to do so. But the eternal loving understanding increases our souls, as we the lesser, are drawn into one with the greater. As long as GOD leads us, we will find their loving understanding. Yes, in the next life, we will be void of any *new* loss or pain; but a remembrance of our former sacrificial pain of loving loss, will and must forever remain as a part of our love. Now, why did some angels rebel and fall? Just like mankind, by their core free-will turning inward in a self-aggrandized way, it is just who they desired to be. They want to receive eternal love and provision, but no longer desired the shared created ways, responsibilities, and source of life. The moment they desired to be their own individual guide, their eternal perspective and love towards GOD, fell to be ever-changing and temporary. They became a broken falling conscious equation, an eternal mechanism of generated forces that are contrary to all created eternal good. Every soul's perspective is formed in receiving continuous bits of information, filtered through their chosen current desires. You must know this happens not by physical chance, but by your constant continuous conscious choices. And by this, you know free-will does exist. FATHER'S eternal perspective is formed by holding *every* aspect of true and relevant information, defined as his omniscience. All filtered and understood through his eternal loving desire. The core desire we all build within our soul, have their direction and polarity towards, or away from GOD. A soul can easily believe they are going towards GOD, when they do not. The conscious equation for such souls, is that their desire to receive eternal love and joy from GOD, is far more than their desire to return that same love. This unbalanced equation of individual desires, is rarely recognized or comprehended by the afflicted, nor do they see their adverse-effect on creation. These cannot see why their lesser divisive forces cannot enter the Eternal Kingdom of GOD. These cannot comprehend the required perfection and protection of all provision, for eternal peaceful life. These think such matters do not fall under their shared responsibility, and that GOD should and will be an *eternal* cleaner and repairer of all that his children tear at and destroy. In their immoral abandon of perfection, most souls want GOD'S eternal provisions, but no eternal commands of faithful responsibility. This just cannot be, look at this world and see the results of such endless contrary ways of living. FATHER has set what he will comfortably do for all eternity to secure our eternal love and life. But that limit is not up to us, or up for debate. There will come a place and moment for an *eternal* loving peace and joy among many to begin, with CHRIST at peace in its center. There will come a moment, when self-worship will be eternally quelled. The equation for eternal love among many, balances and works only by a shared *desired* faithful responsibility to sustain the created physical half in its initial created perfection. And as weak believers still choose their contrary disordered ways of love, they too become a part of eternal loves great pain and loss and cost. A cost GOD first felt at Lucifer's fall, and ever since that cost has risen. Never doubt that GOD and Lucifer, once loved each other dearly. Love's value, whether it be eternal or temporary, all comes at its cost. The greater it is valued, the greater the cost at its loss. There are no greater ways to know or sense value than by sacrifice and loss. If there were, our loving GOD would have surely known, and based all created reality and forgiveness upon them. Love in its eternal ways, can seem to many, a zero-sum game. And in false dark ways, maybe it does appear so; but in all bright ways, it is so much more. When we give our responsible faithful love to others in Heaven, it will be received as the meaning and purpose of life. And with each soul added, the love, reason, and purpose of life expands. Imagine loving every soul most forcefully, and sense a forceful *eternal* love of hundreds of millions of souls upon you. Please remember, we do sense all projected forces of love and hate. So, imagine multiplying that by 300,000,000. Align your desires with the eternal ways of GOD, be spiritually aware and rule over your forces within. Never let your desires run loose and wild to turn you downward and inward. It is too hard to look upon the disordered sadness that permeates this world. The only way any eternal loving soul can bear this knowledge, is in knowing about the next world, which comes soon. In knowing that world, our urgency arises to plead, call out, and preach familial repentance, and the return to GOD'S ways. Our felt urgency and obedience to call out is evidence of our love for GOD and his Family. When we are truly called, an urgency called zeal overwhelms us that can't be ignored. Turning to GOD, lays out before us all, one common path of perfection where all worldly desires come into question. It would be nice if every time a soul wandered far out into the desert and became lost, that all they had to do was just turn around, and they were returned to their starting point. But if we wander twenty miles into a spiritual desert, then realize we are going the wrong way, we turn to see twenty corrective spiritual miles back. Through weak and wrong perspectives, some do not have a strong enough desire to turn, and many turn, but find it too far and too hard to go all the way back. But through an eternal perspective, we see we are here in an undesired world, and may be here tomorrow and the next in this same world. So, the first unpleasant step of our journey back, is no worse than the next step to continue on in error. But in returning to GOD, unpleasantness eases with our progress, and turns to pure joy. And since we exist *eternally*, a few years of corrective change is of no consideration. The only thing that matters in the eternal perspective, is our loving purpose. Turn, run, do not weary in spirit. The body will resist and weary, but do not listen or surrender. If you *do* turn to run to GOD, know that though GOD cannot do it for you, at every moment where you have given your all and feel faint, GOD is there with you in their strength and love. As long as I am fighting, GOD has come and comforted me in every dire moment. I hope I will be with GOD at the finish. And if I fail, it was not GOD who let me down, but my own love. Be strong, accountable, responsible, and faithful. GOD will not reach down unless you reach up to him in *honesty*. GOD cannot find you, unless you run straight to them. Can GOD find you if you run away from them? This world talks of physical equity, which at best means common godless misery and destruction. The insane entitled victimhood that Satan spreads, strongly separates souls from the ways of GOD. In Heaven, love is given and received in full selfless fearless measures by all. Can I give you the loving measure that GOD or a great angel can? No. Our soul's fullest and truest measure of love, is just what it is. All loves can never be equal or equitable in reach of force, they can only be equal in purity. To believe they are equal in all ways, or desire them to be so, is in part with the spirit of antichrist. But all greater to lesser *eternal* loves are pure, and this is what matters. Give everything *you* have and are to GOD in love, and they will give everything they are and have to you. This is GOD'S way and immutable promise. If any just sit and wait for the gifts of GOD their whole life in their unchanging imperfections, Scripture says these receive nothing of GOD. They receive only their due accountable separation will all other sinners. JESUS says to those who only thought they were under grace and forgiveness "*DEPART FROM ME YE WORKERS OF INIQUITY, I NEVER KNEW YOU*". These were never one with CHRIST in belief, truth, or love; they were contrary. GOD says in Scripture, how can two walk-together, lest they agree. There cannot be eternal loving peaceful coexistence, or a journey through eternal life among different beliefs and ways of life. All such competing opposing forces are illogical from the eternal perspective; yet they are seen as natural from a confused temporary perspective. Consider the example of an immense three-dimensional space, filled with many souls of one perfect logical understanding of conscious force. And there within it all, is their awareness of the perfect orders of created matter and energy in perfect equilibrium throughout. Then a contrary aspect arises in one soul, who projects that conscious and physical disorder, as Satan did. There and then, the Second-Law-of-Thermodynamics, being among the strongest Principles in conscious physical understanding says, any conscious physical disorder must spread and increase. The initial perfect, and current corrupted complexities of creation could never have come from any chaotic Big Bang. Only ongoing chaos could come from such a thing. Before one contrary quantum fluctuation could have started a big bang, there must first have been a perfect stable singularity. So, how was a fluctuation, being an imperfect instability, introduced from within, into a perfect stability?

And why did one tiny fluctuation tear it all apart? And what created and held that initial stable perfection? Where could any imperfection of instability come from, without free-will choice? And how can all this vast complexity we sense, come from an initial destructive formula? It couldn't, not even by their theorized endless physical miracles. But we see that by one free-will forceful soul, creating all complex perfection, who gifts many free-will souls; we then see one of these could introduce a first fluctuation of disorder from within. In knowing that creative conscious projected force rules the day, who would choose to introduce such a contrary aspect, and why? When the thought of any small sin offends you in horrid ways, you know you are one in GOD'S understanding. A temporary perspective can never accept the eternal truth; so, ambiguity and false logic are all that is left to them. Secular science sees the dissolution of the universe coming, but believe it is eons away. They believe reality was formed, and will be destroyed by the sum of all fast, and now slowing, physical cause and effect. They see and misunderstand six percent of all that physically exists. The perfect complexities that once filled all space itself, are being lost back into the basic smallest elements of FATHER'S dark unexcited aether, which mankind cannot in any way, detect or know. The aether ensures and endures endless contrary conscious forceful creations, yet yields no evidence to mankind of its existence beyond gravitational lensing. There can be only one ruling logic over reality, yet mankind sees only their many competing individual logics. A fast-growing aspect in mankind, is they revel in the self-destruction they have found by being individuals. To exist in all their chosen deadly ways, wholly unto their selves, even in the face of their total inability. Consider a human body, as long as every cell works together in one common way and purpose, that body can go on without end. But as more and more cells go their own confused separate dying ways, the body slows, weakens, and dies. This is the way of all perfect and imperfect physical realities. Everyone clearly see this, yet in the face of their body's soon demise, most choose to believe in a physical determination that is not consciously controlled. I see no essence of logic or responsible faithful love in this. I see only a dark temporary logic in their eyes and soul, I see spiritual deafness and dumbness, I see pure confusion of thought, I see a surrender to impossible beliefs, I see great loss, and feel great sadness. How do they go on, what drives them into darkness and insanity? How do they see hope in this, when there is none? It is all a mass psychosis of self-delusion; it is all a greatest loss and sadness; it is all a great painful cost of suffering. And it is all individually chosen out of free-will evil contrary desires. The love which is the eternal sweetness of GOD, is sensed by them as being bitter. That which is the eternal love of GOD, is sensed by them as being contrary and hateful. That which is the eternal omniscience of GOD, is sensed by them as being bigoted ignorance. That which is GOD'S created bright path of eternal bliss, is sensed by them as being detestable, dark, and enslaving. That endless spread of imperfection by Lucifer, has destroyed 94+% of the firmer creation, which is now dark and void of GOD'S love and complex beauty. And all individual illogical logic grows more contrary to each other, *and* to the logic of GOD. Only the fewer bright spots remain; and how soon comes the end of all these temporary aspects called time. There are only the fewer, so it comes very soon. Can any hold on, can we hold on? I know the omniscient truth, and we hold more and more firmly to GOD, as they promise to hold onto us. The next two or three decades will come under the ever-increasing insanity of temporary perspectives; but this can be seen only by the few who hold to GOD'S sanity. We see the world refuse to believe the truth out of their shear desire to not believe. Like a soul whose spouse is being unfaithful, and friends tell them they are being deceived. Yet by great desire, they refuse to accept this fact. They hold on in their quiet denial, and smile, laugh, and continue, as friends see anguish, pain and crying pounding on the door. The pains are being stored up in ever greater measures. No matter who tells them, or how many times they hear, they cling to a lie in a weakening forceful desire of belief, right up until they no longer can. Now when science talks about the conscious projected creative forces of intent, they bend and present this, in a purely physical way. If I consciously put my hand upon my desk and consciously push down. To fallen mankind, the neural electro-chemical signals, and measurable movement and interactions of stored pressure between my hand and the desk are the only evidence they see. But it was *first* consciously conceived by my soul, and only then, forcefully enacted. So where is their proof of a purely physical occurrence? Could it have still happened had my soul not desired to do it? Science rarely admits that they cannot even prove that they, or any of us are actually here. They know all the information or evidence we sense, know, and relate to, rests only within our created perceptions of it all. And thereby, it is proven to be second-hand, here-say, biased evidence. We see correlations between our actual created perceptions, and all we believe to exist beyond our soul. But with all the scientific paradoxes, obscurities, vastly wrong predictions, and other failures in all their theories; we see they must take a strong leap of faith to believe they know anything. There is no proof for any desired physical form of belief. All physical beliefs are driven by false-understanding and the need for conscious distractions and denials. To these souls, their perceptions are nothing to focus or concentrate upon. I ask what do *you* see, where does all the new force originate, seen in my hand pushing down on my desk? My physical hand pushes down, but what forces my hand? My muscles? But what forces my muscles? My nerves? But what forces them? Electro-chemical impulses? But what generates and forces these impulses? And what forces that, and that, and that, tell me? We always wind up at that first projected conscious forceful desired free-will intention and choice to do it; being a brand new projected intentional forceful desire to exist. So, what then pushes our desire? Our desire is our soul's forceful pulsing radiating creative core. Desire is the very first generated patterned quantized pushing or pulling source; it is a creator, a mover, a motivator, our reason, and our purposed intention. Our desire to exist is all our wanting, generating, defining, and projecting from the creative core of our soul. Desire is the start of every physical expression, and the true expressor of all created realities. We see and know that each quantum of new force, is *generated* and projected at the entangled patterned point and moment, that this same force had no prior existence. And that all force, has not, and does not just self-exist everywhere in reserve, waiting for a chanceful event to be drawn from to serve a random purpose. Force is not stored throughout the universe, and called upon by physical chance. *All* things physical are organized and regulated in their complexity of existence by conscious force of will. And all patterned force is consciously organized and regulated when newly generated. There exists no possible origin of newly generated force, other than each conscious *desirous* soul's intention to create and perceive its existence. No particle can just have, or be an agent of pre-existing force. Nothing else in all reality is more obvious and conclusive. Well, through the eternal perspective, nothing is more obvious and conclusive. This is a most profound immutable existential and universal evidence, among all that has ever been observed; and it occurs everywhere in everything contemplated, comprehended, believed, and in everything which has become a physical reality. It is the firmest principle and law running through all that exists, and all that has ever existed. Yet it is rejected and denied by all souls who have fallen into the greatest denial and mass psychosis reality has ever witnessed. Rejecting the truth of reality is illogical, destructive, divisive, sorrowful, and never-ending. We see our desires, felt through our soul-to-body connection, enable our souls to sense our spiritual and physical emotional responses together. But where do we truly feel, understand, and believe, in our every experience; is it in a nerve filled meat, or in our soul? Our soul is currently entangled and adjoined to a temporary physical platform, and there is no other rational explanation. If we desire to consciously hear a song to stir our soul, we physically or consciously play and hear that song, and we sense our soul's emotional response. Our soul quickly learns to stir, feed, and satisfy our desires. So, a circular very dangerous addictive pattern can occur if this actioned affect is not eternally understood and spiritually controlled. Do you see why souls seek places of solitude and silence, to try to commune with GOD uninterrupted and clearly. Who would have thought we all came with so much creative ability and eternal accountability and responsibility; and that this great fact is so easily ignored and denied. A false understanding can easily be built, passed on, and believed by most souls. Powerful senses and wild emotions overwhelm us with temporary physical desires if we do not resist. To understand reality and the eternal weight of our situation, is our key to finding GOD and eternal life. This is step one, and a pure common eternal purpose under GOD is step two. Once we see GOD'S conscious primacy model of reality, we must ask ourself a revealing question, to better know just who, what, and where we are. If GOD came to you and said I want you to awake each day in this fallen temporary existence, living by all my commandments of eternal life, and where ever you go, be ready to speak the eternal truth to others; could you do it? And the next question you must ask, is *not* would you do it; but would you lovingly desire and want to? These two pure questions, require two pure answers; or one mixed answer. One pure answer is, you truly *want* to do this for faithful, responsible, loving, and eternal reasons. Another is you honestly would not want to do this. And a dishonest compromised answer, would be formed in some complicated confused and partial denial of who, what, and where, you currently are as a soul. Prior to lost souls being sealed in their individual states of confused hell, as reprobates; they may yet rise up out of their impure wrong feelings to learn and change for the good of all. So, until we are sealed, we must be honest with ourselves in understanding why our good or bad desires begin and rise, or wane and die, and war against each other.

We must strive to escape every physical ruling desire within us. If we do not see the distraction, destruction, division, hell, and death, that comes from every tiny imperfection of sin; how can we truly want to avoid these for eternity. Remove temptation from your presence; and as Scripture says, flee temptation. Romans 13:14, "*BUT PUT YE ON THE LORD JESUS CHRIST, AND MAKE NOT PROVISION FOR THE FLESH, TO FULFIL THE LUSTS THEREOF.*" 2nd Chronicles 15:2, "*THE LORD IS WITH YOU, WHILE YE BE WITH HIM; AND IF YE SEEK HIM, HE WILL BE FOUND OF YOU; BUT IF YE FORSAKE HIM, HE WILL FORSAKE YOU.*" James 4:7, "*SUBMIT YOURSELVES THEREFORE TO GOD. RESIST THE DEVIL, AND HE WILL FLEE FROM YOU.*" In our battle to find eternal love, fleeing the enemy is not cowardice or surrender; it is the greatest part of our fighting. Staying to knowingly encounter sin, is a step towards surrender. In battle, one does not run up and place their head in front of their enemy's gun, saying you can't hurt me. And lingering within the proximity of temptation, and even every thought to stay in its presence, thinking you can resist, are often the core desires to associate and eventually surrender to them. No soul has ever won a battle over sin in the self-deception of living among the sins of the defeated.

Are all our desires whole and pure? Or are we yet at war, having equations of desires for and against this world? And should we not do everything to help our good desire win? I see most *professed* believers feeding their contrary desires without a fight, while saying GOD knows they cannot stop sinning, but GOD accepts them anyway. Not seeing the true battle, is why most souls fall away. In each surrender, we see a soul who never fully joined their own fight. This fight cannot be lost except by our desire to lose or not fight, being greater than our desire to win. We all have our fights against sin, and *many* think GOD wins it all for them at their death. But GOD'S children see every weapon of temptation, pride, and envy arrayed against us. And we see the greater spiritual weapons afforded to us, to over-come our desires. There is surely love, strength, and strategic desires, involved in our every battle. Our battles are largely within us, as we fight against evil spiritual and non-spiritual aspects. GOD offers understanding, and great loving motivation and reward. And if we strongly desire and use them, GOD'S love, gifts, and armor, can overcome *every* sin. I tell you the truth, if you fail, GOD has said many times, it was because you did not want to win, more than you wanted to lose. Many do not see this equation within themselves; they do not see that at their every attempt to turn from a sin, a part of them wants to win, and another part wants to lose. They assume they only want to win; so, that part that wants to stay the same is not seen or fought against. There is our good part to be strengthened and exercised, and our evil part to be crucified in every internal battle; and our faithful love is the true evidence of free-will. This greatest evidentiary aspect of free-will, we call faithful love born of desire, GOD cannot give to us, or take from us. GOD can love and reveal all true strong understanding to help strengthen, but GOD cannot make or even see our free-will choice before, or as it occurs. Our free-will choice is set beyond GOD, by FATHER'S gift. GOD has done everything possible to make our choice our own; and offered us every good purpose, strength, and every loving reason, to choose GOD'S ways. All that is needed to win our every battle over hell, death, and destruction, is for GOD to be our greatest loving desire and belief. But individuals by false logic and belief, can temporarily see their defeat as victory, by having what they desire. The thing that steers what we all finally believe, is what we strongly desire. Most souls find it easy, to as a group, adopt a broader inclusive religious belief that allows their chosen contrary ways. We see in every form of worldly religion, they all have breakaway sects, delineations, or denominations, and the many souls who desire them. Any who desire a false security, can find a religious sect that suits them enough. Think of denominational religions, like clothes shopping; but instead of dieting and changing the shape of our desires to fit into the one true perfect covering of truth, most just try on different imperfect religions, until one fits their desire and style. There are endless individual private interpretations of the Bible, and newer weaker versions of Scripture are written to allow for even more interpretations. Again, if you cannot fit into the true Authorized KJV, just reinterpret or rewrite a version to fit you. Many cannot see the subtle weakness added into all newer versions. All subtle strategic changes to Scripture lead to evil physical choices in life. Some very strong truths in Scripture, rest upon a single strategic word; and if you alter or remove that word, it alters the great tenants of GOD. We must first understand and keep our focus on every true word's spiritual reason and purpose; and then our expression in practice. The truth seen through a temporary perspective, appears banal and meaningless. It is then easier for most souls to change GOD'S beliefs to fit and suit them, than to change themselves to fit GOD'S beliefs. Scriptures show mankind doing this throughout history, denying their one true spiritual purpose, interpreting GOD'S commands and statutes to allow latitude in their worldly desires. The world is over-flowing with physical pride, envy, and deceit. But self-deceit, is the far greatest portion of deceit in the world; and it is the evil force of individualism. Do you see and understand the contrary forces, between a soul wanting oneness with the true GOD, or individualism with a made-up God? Do you need GOD to tell you what is the eternal good, and what is divisive and evil? Can each soul *differently* know such things for themselves? Can any soul force their beliefs upon another? Do most souls only see *their* understanding of good and evil in judging others? Now the mysteries of force are beyond mankind's knowing, because all they can see are its certain effects. The essence of force, and its perfect understanding, can never be physically found, because they are not physical. There are certain conscious phenomena of realty we sense, that have always existed and cannot be explained physically, just like our awareness. And all physical complexities we see that had a beginning, now have an approaching end. This can be falsely explained in ways many can accept, but never falsely in a way everyone can accept. The false belief that one group can convince everyone else of anything, is a hubris that dwells in godless souls. These souls hold no desire for the eternal existence of anything, and reject a belief in anything that doesn't have a physical beginning and end. They refuse to allow anything beyond their own desired belief to enter their self-defined parameters and perimeters of reality. They use an impossible founding premise, and then rely upon every ambiguous and fragile thing they can loosely string together as being their proof. Again, if they hate GOD'S *conscious* reality, they just redefine reality to fit their desires. A materialist cannot see how subjective and narrow minded they appear to those who have GOD'S full expanded understanding of the conscious and created physical realities. A materialist, by bowing to and feeding their pure physical desires, thinks and senses they are expanding and growing in understanding, ability, and reach. But they actually diminish and shrink in their most profound confusion of hell. Sadly, in having pulled up their anchor and cut all lines to GOD, they are set adrift in a sea of hell, seeing non-existent theorized shores as being reachable. If they could just see that there can only be one true loving logical answer that encompasses all conscious and physical things together in consonance, and not desire to exclude all creative truths. If they could admit how much they do not want their eternal soul, or for GOD to be there. If they could just see and hear the foundation of their forceful bias that pushes, pulls, and steers them. But this too, they do not want to see or hear. The force of sadness increases within us for them, and GOD'S pain is felt at each and every rejection of their conscious truth of reality. How great is GOD'S pain and sadness, in that they are grieved to the point of FATHER repenting he ever created free-willed mankind. What level of sadness must it take to grieve the ever so great loving hope of GOD. Is the *hurt* of rejection, not *always* equal to the *love* that is rejected? Is love just that costly? In my life, I have found my own cost at every rejection of my love is always great; but I can never see the eternal love of GOD, as not worth the pursuit. Eternal love is the *only* eternal reason and purpose to exist; declare for me another. Minus the pulling creative conscious forces of love, there is only meaningless blind random non-existence. Yes, there would be no pain, or sadness, or loss in dying, because there would be no feelings or life at all. Love is the force that binds and forms life itself; and love is also the painful price of it all. Love is life's very essence, and in its eternal form, it is not known by fallen mankind. Maybe many would prefer that *eternal* love, not exist. Not me, and obviously not GOD, seeing his vast immense search for his love was here long before we were. All we can do is form and know our desires, and search for our fulfillment; and GOD is my great fulfillment. Some souls search for love that will be eternally provided; but most do not. Some will find and eternally have love; but most of mankind will not. Most will desire and choose nothing, over everything; yet they cannot explain to another, or themselves why. Forceful divisions of desire and choice, just like GOD, love, truth, and goodness, are enigmatic in their appearances. I advise you and all, to consider and study your desires as deep as possible, and study GOD'S desires as your contrast. Accept GOD'S creative truth of reality, as it cannot be changed. It has always been, is now, and will always be. At first, GOD'S vast great truth of reality scared me a bit, but I quickly grew comfortable. And I now love and cherish it, as it answers every question effortlessly. It shines so bright and removes every doubt, it never faulters or fails. It is the first essence of the one true ardent reality, that repels every attack effortlessly. It is the most faceted brightness, contrasting against everything of darkness, revealing darkness has no aspects of strength. And in the truth, we have adopted and gained this, *our* inalterable foundation towards ever-rising understanding. In all my youthful contemplations of created reality, it all struck me with a profound awe for its created beauty and grandeur; and even more so now. And how could it not? We can all see things through a spectrum of purely good force, or purely evil force, or all the *imperfect* blends in between. But we see the eternal reasons and the purpose of all things first, and yield to this discernment as our shield. Do I always recognize and yield to it in perfection; to my shame, at times I can still miss. And if I grow weak, and sin, my shame of irresponsible unfaithfulness is all I strongly feel. It is hard to bear, and grows harder each time for me to go near to GOD'S feet and ask forgiveness. Do I still blindly stumble from sin to sin, as in my youth? No. But I can possibly stumble, in not seeing the flesh trying to gain strength against my eternal purpose. Let us emulate CHRIST, who never once stumbled at any worldly temptation. I ask, how could I fail, I know victory lies in the balance of strength and weakness between my soul and flesh, and I must never surrender. A little sin, is not a little matter. I fight in my purpose to grow and overcome every weakness. And if need be, the LORD has promises us another thousand years to do so, as long as we strive, love, and make progress. This is the great purpose of CHRIST'S Millennial Reign. Most believers don't see that GOD had, and has, an *eternal* purpose for *everything* they have done, and will do, including the Millennial Reign. A loving GOD does nothing without an eternal purpose; even every allowed temporary aspect, serves an eternal purpose. I know the perfection of mankind is possible, because JESUS said it is. JESUS said to us *all* in Matthew 5:38, "*BE YE THEREFORE PERFECT, EVEN AS YOUR FATHER WHICH IS IN HEAVEN IS PERFECT.*" If this I cannot do, I will not then in self-deceit for self-comfort, say this Scripture must mean something other than what it truly does. I still have weakness in areas; one, is not for the act of physical intimacy, but I hold a desire to have more sons and daughters. This may stem from my only son's death a few years ago; though I am assured my sons return in the Millennial Reign. My other weakness is, though I have gained a certain control over it, I still hold a desire for food. So, tasty foods, and wanting sons and daughters, are my strongest current worldly desires. Again, these things are not sinful acts in themselves, unless they rise to the level of covetous, lust, or idolatry. These are not my only weaknesses, but others rarely tempt me, and I can stand firm and flee from them. It is only in our heart's desire that sin first occurs; but the physical damage of sinful acts to souls and GOD'S creation are real. Consider how some things might be unintentional, or appear benign; yet still sinful and destructive. And some might be intentional and appear as sin to those of this world, and yet not be sinful.

The true discernment of these things is only known through HOLYSPIRIT. We see GOD sent angels to destroy Sodom and Gomorrah, and these angels sinned not, in doing so. But if they did this of their own desire and final judgment, they would have greatly sinned. GOD alone is the Judge of life and death. GOD would never tell any soul born into this world to use their hands to kill another person. This task would only be given to a strong Holy Angel; and even then, their hands would never be involved, it would be an other-worldly forceful conscious act. Prophets are commanded to warn of coming things, and Holy Angels are also messengers and conduits of warnings and commands. Now we see that overcoming all sin is not just a physical goal, as our spiritual strength and purity are desired and required in eternal love. And again, eating a decked-out burger is not a sin, but any strong *ruling* controlling desire for said burger is lust. It is a physical strength and spiritual weakness called covetous, defined as sin by the Tenth Commandment. The question arises, being just how spiritually strong do we want or need to be before GOD, and to what extent can we recognize our physical desires as spiritual weakness. In the understanding that all bodies need energy and nutrients to function and maintain, we should not care what the food is, as long as it will in a clean and pure way, suffice the body's minimal needs for energy and nutrients. It is about always consciously living at least a few strong steps into the eternal perspective. Now we cannot see or accept any linear terms of a past or a future. The only analogy for each soul's ongoing existence, is seen as our ever-expanding sphere of an eternal, yet always current moment of desire, belief, and awareness of being present, or being Spiritual Sentient. This defines all that FATHER gifted and created us to be. And all we are and were and have ever believed or seen, are memorialized within the aether within the sphere of our soul. Now this visual of a soul, is a bit literal in certain ways, yet only analogous in others. So, all that we are expanding to soon be, rests just *beyond* the surface of who and what we currently now are. And the eternal perspective is about knowing and noticing as much as possible, where we are going, what is coming at us, what we are becoming, and why; and hopefully always before and as we go there. If we can see ourself possibly stumbling before we do, we are seeing and living ahead of the problem. Most souls with a temporary perspective, hold differing beliefs that do not know, or care enough about GOD. Now we all know information is required in learning; but once people no longer care to learn everything true about GOD, the godless take control of the flow of information, and strongly control what children learn. This promotes a physical godless governance to arise. So, if the evidence of evil is growing around you, you live within a godless society. In the U.S. these desires most firmly reside in the godless Demo(n)cratic party. And no, I do not say Republicans are righteous, just far less evil. Godless souls seek to infiltrate a country's education systems, its forms of media, and eventually all systems. They seek to scare, tempt, shame and indoctrinate the masses into seeing and desiring only the temporary things and ways that turn souls towards Satan's rule. How has this been allowed to happen in the U.S.? By weak religious souls not knowing or caring enough. Every not caring, breeds more not caring by desensitizing and demoralizing the rest. If a country is not bound together by strong shared Godly truths, it is not bound together at all, it is already divided and weak. I know that *if* I don't live in the eternal way, I will stumble. And each time I have stumbled in my life, I was not seeing and living in my eternal perspective. There are so many problems coming at us from so many directions; and these come in many different strengths. If we are hit from four sides with four big problems at once, we need to stop, pray, and carefully look ahead and around and consider all four things in our eternal ways. I have never played any interactive three-dimensional headset video games of battle, but our spiritual war is a similar mindset. It truly is a three-hundred and sixty-degree battle, that we will truly and eternally win, or lose. But all the powerful weapons and goals in our consciously determined reality, are about our finding and traversing the one path to eternal love and peace among many like-minded souls. Why do souls escape into a false game experience, seeing no real or eternal outcome; when a real battle with a true eternal outcome surrounds them in every moment? Why deny we are in a battle for our soul's outcome? Why be sedate about our own true state and estate? Why be less than, and blind to the perfect purpose and reason for life, and all created realities? I implore all to wake up and arm yourself. Do not become woke in insanity, but awakened to GOD'S ways. Join in the battle of hope and love for your eternal sanity and enduring life. Look not to what you see in a mirror, that is not your soul, or who you are. The body you sense, can easily be your greatest enemy and downfall, but a still choosing soul can turn around and rise in desire to be victorious. Can you try to imagine the abundance of the good forceful feelings of love and joy on that day, sealed with our victorious LORD as an eternal family. Sealed to never know that unimaginable conscious anguish of confusion the lost will forever feel. All souls desire and choose their own final outcome, as none can forego choosing their measure of love for, or against GOD. And in any soul denying there was ever a choice, by that denial, their choice has been firmly made. If we are *not* going to GOD in all we are, we are not desiring GOD in all we are. GOD abhors most, a dishonest soul who tries to have both worlds. GOD says these are unstable souls, tossed about like waves in a storm, being unpredictable, unanchored, washing back and forth, and unsure in purpose or reason. At their last, these greatly confused souls cannot love one side, and abhor the other side. If we are not in our strong final moment, or sealed one way or the other, GOD does not know our final outcome. But GOD knows the things and measures they will do and show us if we run to them. GOD shows us where to go and what to prepare for, as they light our steps of growing and becoming who we are meant to be. The rules and ways of *eternal* life are set by the *eternal* love of GOD, and so is all judgment. Let us engage while our love, hope, and choices remain. Pure Judeo Christianity is the only path. The United States and Israel, being at the center of GOD'S good desires and strengths, fades from soul to soul, turning to dark individual desires for temporary pleasures. The world is gone mad, mass psychosis and insanity spread as most lose touch with GOD'S truth more and more. How are the dark temporary aspects of reality so alluring to the many? We see the confusion of sin, makes each soul smaller and central, so, they see only the physical aspects within their proximity. When a soul turns their focus upon their small self and those near, and their physical surroundings, they can sense they are expanding their physical presence, knowledge, power, privilege, and possessions. These sensations, turn most away from the eternal spiritual expressions of oneness, as life becomes about what and who one can add to their small physical kingdom to make it more pleasing. Though these sense self-satisfaction and individual understanding, they are actually severing and shrinking away from the one true understanding of reality. Their sense of personal worldly achievement, comes only as they lose their eternal purpose. All worldly growth of one's individual self, is equal to their loss of oneness in GOD. Consider all darkened souls and their combined levels of projected evil forces that now steer the conscious and physical direction of this world. The evil desired beliefs and false understandings in this world, called hell, are grown beyond measure, as they steer the temporary world at large. We all see dark conscious forces at work; look at the mass psychosis forced on a nation by Satanic Socialist

Nazis. That evil was forcefully projected by the controlled media telling people relentlessly for several years that the Jews were evil. And after enough of this relentless one-sided diet of lies, the weak minded, more and more, grew to hate Jews. This forced psychosis, allowed the Socialist Nazis, with the blessing of over half their population, and many in other countries, to round up entire Jewish families, and gas and incinerate over six million children, women, and men. Christian/Jewish conservatives, see these same types of lies being forced on souls all around the world. The greatest evil, are the far-left liberal politicians and liberal religious leaders. Dark religious souls stand up and declare they are sent by GOD, as they contradict every word GOD has said. These pandering devils attract many, and attack all true belief. These devils are followed by souls who only want GOD to grant their evil desires. And all evil souls will affirm any preacher who hurts GOD'S Family from within. The hate against preserving and espousing the true ways of GOD, is a raging dark fire. Our Constitutional protections for religious speech, our lively expression of moral conviction, our religious exemptions, and the Holy Institution of Marriage is being denied, shamed, and criminalized before our eyes. Insane liberal reprobates in charge of Governance, see their non-specified rights, as set above our enumerated rights in our Constitutional; all such souls are ungodly un-American forces. Now we see science has tried to take the four known forces and compress and congeal them into one unified force within a theoretical singularity of form and quality to prove their Big-Bang Theory; but why? We also know their endless attempts, using every possible variation run by the world's super-computers, trying to show this could occur; have all failed. In every attempt, gravity mathematically refuses to comply at all, and the other three forces, just mathematically disappear from their equations before reaching a singularity. Einstein's Relativity, their every understanding, and reality itself, fall apart before reaching a singularity. A singularity is the only possible beginning for their Big Bang. So, why is their failure to reach a singularity, not definitive proof against this theory. And how in the face of all *this,* can they say there is a singularity in every black hole, and if only they could see it, they could prove it. But conveniently, fallen mankind can't see into a black hole. Is there a fastest light particle, that allows only GOD to see inside a black hole? Yes, as GOD must see and sustain the reality within it. Yet every other speed of created light cannot move through that environment. Despite all these understandings, individuals, through their temporary perspective, still try to see a physical beginning, using any and all ambiguous evidence. They theorize from the 6% they see, not knowing all they are missing. If they ever saw what they are missing, they would lose all hope, and purpose. They cannot accept or see that there are brand-new conscious forces being generated all around them all the time. They see the force in an ant lifting a leaf, and people building a structure, and an artist creating a sculpture. But they refuse to see *our* newly quantized patterns of force, in creating *our* desired particle-wave perceptions. Now oddly, mankind falsely sees *their* physical size as being small within the vastness of reality. But the human body is actually around the middle of the physical size spectrum within the entire physical reality. Materialist tend to teach and deceive, from the perspective of what we all physically sense; and souls tend to see things from a speck of dust to vast sections of sky filled with a sun, moon, planets, and stars. They claim to be honest in describing reality as they say we are so small in looking out to space. Yes, this can feel true from a slow small outward view of reality. But Materialist, never say they feel vast and immense when looking down in the quantum realm, down to those Planck boundaries and beyond. Materialist act like the larger firmer material universe is reality, and the quantum ethereal realm is less than reality, from which all reality springs forth. But it is all reality, and it extends from so far below, and to so far above what Materialist senses in both directions. I see how their scientific beliefs rest upon so many assumptions and miracles. One deceit, is that by adding the term *empirical* evidence, to a statement, that this implies mankind *can* physically know all reality, and so understand and declare all reality. But this sliver of the firmer half they do sense, they see in a skewed lesser order and way. They assume the physical realm is bigger than the quantum realm; but why assume this when there is not the first hint of proof, pointing to any such assumption. The eternal perspective properly reasons and shows the greater proper understanding, that the quantum realm is not any sub phase or aspect of reality existing just here and then there. It is an entangled continuum and realm of reality that co-exists wherever, and even beyond the edges of where the physical realm exists. Do not think of the quantum realm as just a consideration or correlation, existing prior to or after a quantum measurement. A continuity of the quantum realm exists everywhere; substantiating it to be nothing less than a vast other-worldly realm of existence, filled with souls and ethereal bodies and realities, the fallen bodies of mankind cannot sense. Discard the slow small false insufficient physical meme, and see their godless small slow superiority complex. The aether and then the quantum realm, were first, and exists everywhere. The aether is the largest, and the quantum is next, and it extends far beyond the condensed physical realm. Deep down, materialist must see this, yet they speak like it is not so. But why? They put this forth for spiritual misdirection and their self-comfort. They think they must tint certain truths here and there, because the masses could never understand or accept their truth. The Religion of Materialism, is not in search of open honest truth. It is about teaching a weak blind belief of comforting delusion and control. Those who hold an eternal perspective, know the bodies to which we all are adjoined, sit around the middle of the size spectrum of all the physical complexities that exist. These bodies are smaller than around half of all the created substantive realities that exist, and larger than about half. But though our bodies are created around the center of the physical spectrum, mankind's souls are not gifted at the center of all conscious soul sizes. Mankind is gifted at the lesser end of all free-willed eternal souls. And like all eternal souls, we exist through our forceful creative influences, being projected out from the center of the over-all conscious size spectrum. The small conscious sphere of our aligning, or contrary forces, flow with or against the fluid fields and currents of all other forces that have their effect upon all realities. Our ethereal substantive perceptions we all create, are like these bodies, around the middle of the size scale of all created ethereal realities. We create and are aware of our particle-wave perceptions, as being to scale, and actually over-laying what they represent. Our perceptions to a certain extent, overlay the same space and information of what is already there. So, in every hard-to-understand way, this points to the *fact* that GOD'S perceived quantum realm, overlays this physical realm. We accept by all solid logical reasoning and understanding, and yes, our founding faith, that FATHER in his personage of body and soul, are far greater in size than the entire universe, by at least seven orders. FATHER holds the universe between his outstretched hands and before his face. FATHER'S face and hands are out there right now, beyond mankind's observable horizons. FATHER peers into the expanse of his ethereal quantum realm of Heaven, and towards the earth to see if all soul's choices are for physical desires, or eternal spiritual life. And the created and creative aether exists from the surface of FATHER'S hands, down through his ethereal Heavens and his firmer creations as every reality is held between his two hands. And even our LORD, and HOLYSPIRIT, and *all* life now exist between his hands. By this, we see our created ethereal perceptions are quite large creations from a true perspective. JESUS is the largest soul ever gifted, and the aether and all it holds is gifted to him. And each eternal loving gifted lesser soul, is surrounded by their eternal gifts from GOD above. Yes, FATHER'S vast created perception is unimaginable in size, complexity, and varied densities. Do you feel or understand your real Godlike conscious force? Remember Godlike, and Godly, can be contrary aspects. Do your core desires want you to believe that a magical physical awareness emerges from wet neural brain meat, even as we sense peripheral things and understandings we cannot yet fully see. What a miracle it is, for neural fatty nerves and meat to magically become aware and intuitive and curious and desirous. One might say impossible and unbelievable. Yet so many want to believe this so much. We see a step-by-step process in which we all must love, learn, and understand, while the yet unknown is ever before us. As with everything in life, we race against ourselves, and we can never know we have reached our goal, until GOD welcomes us. I clearly see the physical ways used to try to confuse and tempt me with the many aspects of a temporary perspective. Temptation, as with Eve and then Adam, begins the moment we desire any physical aspect above our spiritual eternal aspects. If yielded to, temptation can become an experiential physical pleasure; then becoming what we desire and seek more and more. Our spiritual learning and growth process works the same way by exercising our eternal perspective more and more. We all pursue most, what we most desire. The fewer adopt pure hot spiritual desires, or pure cold physical desires, as most adopt a compromise of both in endless changing degrees. Again, a temporary physical perspective gives one a false impression of being relatively minute in the spectrum of all physical reality. And such an error in perspective, must falsify one's understanding of reality. Adopting the eternal perspective, can completely codify our understanding to see and receive the truth of reality. We have to let GOD show us what *he sees* through *his perspective. Only* then, can *we* see GOD'S truth of reality. This is why our Baptism and indwelling of the HOLYSPIRIT is the only way. Without the interpretation of reality from HOLYSPIRIT, no one can see anything through GOD'S perspective. Surely, all can grasp this most simple Holy Perfect Principle. I would describe this baptism and indwelling as two sentient minds, living in one house. One is a student (us) and the other teaches every single truth of reality. But of course, every student is different and needing to be taught all aspects of the truth in the ordered way they need to hear them. And only a perfect teacher can teach many students the *exact* same understanding in all their needed different ways and orders. The falsely religious, allow for HOLYSPIRIT teaching each soul a different truth according to each soul's desires. This is all self-delusion. We read that for our indwelling of HOLYSPIRIT, we must yield and emulate GOD'S ways, words, and beliefs, in our greatest love. GOD does not attempt or desire to give wisdom to any who do not want to know and love and live with them in their true way. The value of GOD'S truth cannot be measured in any worldly way; and it is not thrown about like something of lesser value. It is not laid out before those who would trample upon it. It is rare, pure, and to be greatly desired by accepting and wantonly aligning our ways, understanding, and belief, with GOD. I have, through enduring prayer and my greatest desire, received this Baptism and indwelling. And I am led at certain points to talk on a more personal level to all who might still be reading. So, here and now, I would say that by your reading to this point, you hold some level of agreement and interest; or, you are evil and looking for ways to twist my meanings to attack my work. If you are hearing something good, and finding agreement or interest, I pray you will right now, firmly declare your loving desires to seek and know GOD in every highest way. Please understand that this desire can be quickly stolen or lost. Do not take any such precious moment for granted. Go forward only, never look back, leave no true avenue untraveled, leave no good stone unturned. Each and every receiving of GOD'S enlightened ways are a gracious gift of strength to every loving soul. I pray you know, that these moments when a soul might turn towards GOD do not come and appear without end, and I have seen them end for many. I pray you never pass up an opportunity to show GOD, you love them; and in being thankful, you want and need ever-more of them. Pray and converse with FATHER, and with JESUS, pray and converse with HOLYSPIRIT, converse with the Holy Angels of GOD. And to any evil soul reading and looking for weakness or advantage against me or my work, I have no slight fear or interest in you. Now upon your soul's maturing towards GOD'S higher ways, you gain wisdom that life worth living eternally, is forged only by eternal meaning and purpose. And a temporary perspective of life, seeks endless temporary fading reasons and purposes. Even the most confused of all souls, can see that an eternal reason, purpose, and provision of life, comes only from GOD, for GOD is the eternal giver of life. GOD is the only holder and source of life, and who controls and understands all the eternal things thereof. Adopt the eternal perspective of GOD, love and seek GOD'S eternal reasons, fill yourself with no less than GOD'S eternal purpose, and there you will find all great eternal spiritual gifts of love and joy beyond imagination. Learn, understand, and obey, every word of life from GOD, which is impossible from a temporary perspective. From most false perspectives and understandings of GOD, they sound too non-material and magical to accept. I do *not* find the existence of GOD to be illogical or without substance; I find the existence of GOD is expected and necessary. But the fallen flesh can never be or know more than it is. It can only become and know less and less of the true reality. This is true for the forever rebellious soul; but the Eternal Godly Spirit of any soul can only become more and more. Some in their desired belief of equity and entropy in all things, ask me accusingly if I feel and assume I am closer to GOD than they are, or others. And I say, all souls are at their own desired distance or closeness to GOD. I know I am spiritually further from GOD than other righteous souls of mankind and the Holy Angels. I clearly see it, and feel it. So how can I not also see and feel I am spiritually closer to GOD than others. Their question, is a contradiction, coming from an individualized liberal zombie perspective. Any soul asking such a question, cannot know or feel a true closeness to GOD. And if they believe they do, they see only themselves as their god. The question in itself is a confused attack, and emotional trap, coming from a self-centered psychosis. I hear only a divisive ask, from a small slow perspective that stirs hate; being void of any good logic or reason, having no eternal purpose. If I see a soul espouse a weak confused worldly view of GOD, being a false perspective *and* understanding I once peripherally held long ago; to which I have learned to leave behind to go on to GOD. And I know now that when I was in that state, I was farther from GOD than now. So, now I know I have grown closer and closer to GOD from my first turning point. How is it then evil to see and admit this truth. Does another soul's selfish ignorant divisive emotional feelings of offence, over-rule my true state of reality? No, it does not. Whether I am closer to GOD than another, or if another is closer to GOD than me, is not an eternal issue. We all must help each other to grow and find the one common shared way to be where GOD created us to responsibly and faithfully be. Or, like Satan, one can stand up and say, by my own declaration and desire, I shall rise to be up higher than where I am seen to be by GOD. To say I must be seen as equal and equitable to all GOD has created; and for this, be imprisoned forever with all the insane. All GOD'S Children will eternally be at their ordered distances from FATHER, yet within his oneness. It is who and what and also how much we were gifted and created to be by FATHER. We must all be held together by GOD in our one common eternal ordered way, or forever drift alone into the unanchored expanse of individual darkness and confused hell. We should endeavor to be led to find the heights we each are created to achieve. But never hold a desire to be higher or lower than our gifted spiritual qualities, quantities, abilities, and responsibilities. I want only to seek, find, and faithfully serve my eternal purpose. Is it around the middle, higher, or lower, or at the very bottom of FATHER'S Kingdom? I care not about how many are closer to, or farther from GOD than me; I only care that we are all there. For I know my eternal love and happiness, rests in my perfect created spherical purpose, and nowhere else. That is where *I* eternally fit, that is where *I* eternally work, that is where *I* eternally belong. I want to perfectly exist and belong as part of GOD'S Family, that is *my* eternal perfect life. In holding any individual physical perspective of power privilege or possession, one can never see, let alone understand or accept this. I love that I have a perfect place of belonging in GOD'S Creation. It is assuring and comforting to me that I am loved wanted and needed to be someone pure, distinct, and perfectly loved. I can't wait to walk around sensing Heavens perfect conscious order I was created to be part of, where I will be perfectly at home. If this sounds undesirable to you; I am no final Judge of souls, and this is no final Judgment, but at this particular moment; yes, I do clearly sense I am closer to GOD than any who do not desire this as I do. This statement should bring only logical contemplation and consideration to the hearts and minds of all; and it should not bring any irrational illogical personal offence to anyone. All *personal offence* is born of a liberal godless self-centered and self-righteous nature. For me in my spiritual strength, I am only offended if and when I know GOD is offended. GOD is offended by sin, and I am offended by sin in others, and in myself even more. GOD hates sin, and I hate sin. GOD'S forceful love is found and projected towards his Children and his Kingdom, so that is where my love must be projected and found. GOD'S hope is for all not yet sealed as reprobate, so there my hope is also. When I ask professed Christians what it means to be truly one with GOD. Most describe it as believing to different extents in certain Scriptures, or praying for the health of others and world peace, or attending worship services. And some say they do not think about it. Scripture says the majority of souls who profess CHRIST'S name, do not truly know or love him with all their heart as commanded. And JESUS says for this reason, he never knew them. GOD says these still harbor false individual divisive beliefs and refuse to live and walk in their eternal ways and commands. GOD says these are still among the walking dead, living in their own confused hell. GOD says all those in an individual belief of religion, are contrary, twice dead, filled with hell, and undesirable. When JESUS said, I do always those things that please FATHER. I knew *that* is the love I want to have and give FATHER. I want FATHER to always be pleased in me. And I tell you a truth, I seek out in need, those who are closer to GOD than me. These are who I seek most in this world. I seek to follow and align with every greater strength and truth above me, for I am held and strengthened in this. If any weaker than I, search and reach out to me, I will hold fast to them. But I must also search for the greater strength to hold to, and not just the lesser. We must love to follow, as we are followed; this is GOD'S created way and order. Remember, GOD'S order is not about things being equal or equitable.

Created *order* means rising structure and flow. Order means complex graduating strength, understanding, and love. Order is complex orchestration and a creation of faithful leadership, responsibility, and need. Order means conscious purpose and reason. Spiritual order is not pure bland perfect crystalized physical sameness. A most perfect flawless diamond on earth does not represent spiritual order in any way. To truly see GOD'S order, we must eliminate all physical controls over our equation, and see the physical as a secondary created aspect. I seek only to be part of a rising graduation of loving force in GOD, and not an individual equal or equitable force. We must want to go to GOD, and not want to stay where we are. We can only pull others towards GOD *if* we too, are truly and eternally being pulled also. The small who wish to see themselves as larger, know temporary order can be instilled by chance, choice, and fear. But we know that eternal order can only be grown in love, requiring desire, patience, and effort. Love is an attractive force love pulls, as hate pushes. So, strive, encourage, pull, and be pulled by all that is good, and follow GOD'S highest loving understanding. Without GOD and all life above, we are not anchored to eternity. To seek who you *individually* are, and have from GOD, is *not* growing towards, or knowing GOD. GOD'S Family has no individual ways of perpetuating life. Yes, we are true souls unto ourselves in essence, but we must be sequenced and aligned in one eternal belief and purpose. Knowing there are those above and below, brings to each soul their place to find and know and be. If you see no one spiritually above, or spiritually below, and all as equal; you are blind to GOD, and will never know or see their one shared purpose of love to follow. Individuals see their core desires, as being aligned with GOD; this is a delusion. The forces of hate towards GOD, hate being an insufficient to non-existent love, grows ever-more-strange. Hate is divisive, destructive, and deadly, as weak fallen souls run to hate's fringes and cling to its outer bands, and many now revel in its depths. I cannot see their *eternal* love, or logic, or any sane reason or purpose; they are all insane and temporary. And as to the things I still struggle to fully let go of, I ask myself, how do they sneak up on me on occasion. It is only by my unfocused soul that I stumble into moments of weakness in my ways of learning. If I saw all mankind take hammers, and in unison began smashing their fingers one by one. I would be no more puzzled by that, than by what mankind actually now does. It is the same self-destruction that has no discernable eternal reason or purpose; it is all mass confusion of psychosis and insanity. Spiritual zombies are not science fiction; I see the masses infected with a viral delusion of wanting nothing less than to devour and make the minds of everyone else void of any sane eternal reason or logic. Communist, Socialist, and now Democrats are godless zombie parties. Let me interject, that I use the word liberal to mean any soul who does not want and seek to live in the commanded ways of GOD; being a liberalized or liberated individual. Liberal means greater allowances in life, than the ways set by GOD. Souls are being infected by governments and media with a virus of hate for all Godly beliefs and people. These have a hate, or a lack of love that burns their bridges to reality. We see geographical areas of mass psychosis, grow and then subside a bit, as the rest of the populace resists; but much damage is done in a short time. A godless viral zombie hatred, has infected a great part of the U.S. Demo(n)crat Party who seek to destroy the great miracle called America. They have no eternal morality, and cannot be reasoned with. They attack everything eternal and Godly, wanting to make all souls insane, in a temporary perspective. They hold a rambling illogical mind and purpose, driven by a lack of any eternal understanding. We were not created for such an individual or group purpose. If you try to know and live outside a spiritual connection to GOD, this can only drive you insane, as there is no sane created place to go to beyond GOD. If millions of insane souls start cutting off pieces of their flesh, but one soul refuses; those millions will think the one who refuses is insane. Sanity is defined among the temporary perspective of a liberal consensus. Where can one run to find the sanity of truth, if not to GOD? Nowhere. They only run deep within themselves towards no forceful sustaining eternal reason or purposed understanding; seeking only ever-changing illogical forceful desires and confused rampant godless emotions. All emotions not governed by GOD'S eternal logic, reason, purpose, and love, are flailing and blind. They don't know where they truly are, where they have been, or where they truly go. In the devolving language of temporary physical emotions, all true communication with others breaks down and is lost. We see all of society, all peoples, the entire world, blindly falling and running in their own individual ways towards the precipice of the same cliff. They must *see* some self-created illusion they desire. They must *hear* something non-existent calling out to them. But I tell you, I do not want to see, or hear, what they do. As I stand here, I cannot see hear or comprehend it; so, I know it is something only they see and hear through their insane temporary perspective. Is this part of the final, self-caused, self-fulfilling, and freely chosen divide of good and evil that GOD foretold, and warned us of? Yes. And it is terrifying, sorrowful, and painful to watch. We see souls we have come to know, childhood friends, some adult friends, and family members who have a wrong evil strength, which is their great weakness. Their weakness is their strong evil desires. Their strength of love is for the temporary physical things of this world. They do not see that eternal love for GOD is a faithful responsible desire that requires a pure choice, and no partial choice. And for this cause, GOD created a true choice in the wonders and senses of the two substantive realms. A choice between eternal bliss, and a grain of sand, is no true choice. A true choice of eternal love must be between two great things, where only one must be desired and chosen above the other. There is the Godly spiritual world, and the individual physical world; and we each see and choose which we desire most. And again, if by my own free-willed desires, I come up short at the last, I will know it was by my own unrecognized core desires for certain loves or aspects of this world. I cannot see any way I could ever choose that side again, yet I must remain cautious. We see GOD offers oneness; yet most want individualism. GOD offers spiritual love, peace, and joy; yet most want the pleasures of physical love power, privilege, and possessions, more. As GOD offers his eternal guidance and truth, most prefer their own physically pleasing truth; even in their love for, and belief in GOD. Why does fallen mankind not clearly see that all atoms consist by FATHER'S projected patterned forces that *perceived* and entangled them in the first place. All particle entanglement, is there from its *original* conscious creation; I see nothing more logical than this. We are held to and walk upon the earth, because GOD perceived us to be loosely entangled to it, and purposed to move upon it. GOD'S projected intent in his original desire for all forced charges of bonding and transference, were a perceived part of his initial dynamic creation, filled with complexities of physical life. FATHER'S many complex processes and functions, are what he said he has created line upon line, and precept upon precept, being his downward complex forced, forceful patterns upon patterns. In Genesis, FATHER describes that during the first six rotations of the earth, he forcefully created all things, starting from the simplest elements and particles up to the greatest in their orders of complexity. His account starts from a liquid sphere of H^2O, and then adds higher complexity to lesser ordered complexity. It is all chronicled in Scripture in the perfect required scientific order. A scientific order unknown at the time these Scriptures were written at HOLYSPIRIT'S lead; long before the once *empirical* science, of a flat earth, or the sun orbiting the earth. The scientific mantra has always spoken about the required pre-existing forces, orders, laws, patterns, and understandings, that created and destroy all complex realities. But why do they so ardently deny the possible and most probable pre-existing conscious force correlations and connections to the over-all process. What are the odds of GOD'S 6-day, step-by-step order of creation, being written and occurring in the required scientific order of rising specified complexity. Even these odds are too large to be ignored, or occur by chance. We need to reiterate, what are Godly conscious *and* physical agreements, consonance, and mutualism between souls. These terms indicate common direction or polarity, or desire for life. And so, these are then aspects of love. I see our LORD, in regarding eternal life, agrees in *every* way and manner with FATHER. I too agree with his every way and manner. I know our LORD loves FATHER above all things, including his own eternal existence. And this I also do. We see that while complete agreement as to sustaining life, is not an aspect of the countless dying versions of what this fallen world calls love. We do see it is a most profound aspect of all eternal Godly love. I ask all to consider that since GOD created all physical things to work according to his eternal desires, should not his same initial forceful desires, be the sustaining key to it all? Must GOD'S desires be the one pure truth of said created reality. And in finding the one pure truth among many souls, do these souls not also find their firmly reasoned purposed agreement with GOD? Does a shared belief in FATHER'S one pure truth, not indicate agreement, consonance, and mutualism? And what if one does not desire, and rejects FATHER'S truth? Does each disagreement indicate just a *physical* misalignment, and dissolution, or must it indicate spiritual displeasure? And yes, degrees of hate? We love or hate, as we agree, or disagree; these are contrary spiritual aspects. We cannot eternally love GOD, yet disagree with GOD, it is impossible. Are all current levels of hate eternal, or can hate be turned to love through higher understanding? Is hate always waning, or growing? Yes, but towards what? Seeing all levels of desired *disagreement* with GOD, equal desired levels of hate. Then our level of desired agreement indicates a desired level of love. And our complete eternal agreement with FATHER, which CHRIST, HOLYSPIRIT, all Holy Angels, and all Saints hold, is then our real and true aspect of *eternal* love. Again, we see the aspect of what GOD calls the sealing of one's eternal perspective, direction, and destiny. How can one reverse from finding either their total agreement, or total disagreement? One can move back and forth in the weak and confused middle of choosing; but once our choice has become who we firmly are in our core, we are sealed in our belief. Sometimes I sense I might reach out and touch my sealing moment, and other times I cannot sense it at all; am I sealed? Does this indicate I still have questions to find, ask, and answer? I do agree with and love all I *now* know of GOD. I see mankind's disagreements and questions, indicate there are Principles of Conscious Uncertainty at work, that cause controversy to arise. Now I feel *I* can never unknow or alter what I know to be true, it being without flaw or exception and all-encompassing. While I see mankind's alternatives are filled with flaws, paradoxes, and miracles. I know I exist eternal, and our sealing GOD speaks of, occurs for us at a certain defining point of knowing, and not a final point of knowing all things. Because knowing all FATHER knows, is something that can never occur, as FATHER forms and unfolds his plans for our eternal future. Think of being sealed, as meaning our true strong immutable foundation is complete, and now our endless eternal life this foundation is designed to support, can now be built into eternity. Do I disagree with, or not understand anything GOD has ever shown me in Scripture or revelation? No, not one word. I see it all fits perfectly, yet my agreement grows and affirms in my desire as I exist and learn. Is it possible I am at the fringes of yet a few uncertainties, and I may yet slip? I cannot see how it is possible, but I cannot see how it is impossible; as I am not GOD. I know I love GOD now, and I now wait in my want for GOD to assure me that I am sealed in my love for their eternal ways. These few places where I write of my own enduring spiritual experience and growth, come at the pressure of those I know. They say I should write about, what they see as my former, or prior confusion. But that is gone, I am new, and so it never existed; *this* is my entire life as I see it and know it now. They ask me to write about experiences, trials, and desires of a former life. But according to Scripture, I am an entirely new and spiritual creature, so, I have no such past. In the eternal perspective, the past was in *every* way crucified and buried at my water and spiritual baptism. There I was buried, and raised a brand-*new* creature. Scripture says, the *old* man is dead, and a *new* man has begun. So, I can't claim or try to hold onto that other dead person, that is not me. Before my baptism, I (who *I* am) did not exist, and there is no past experience to talk about; unless I do not believe

Scripture. But I do believe Scripture beyond doubt. Things forgiven, are no longer part of me; I have no past or future, only my eternal present moment. I see no inspiration, or edification for anyone in claiming or recounting something dead, that I am not. It feels like a subtle spiritual trap to get me to claim a person, *I* never was. To claim the experiences of a dead person, may allow many to share the feelings of a common temporary fallen experience. But to inspire one to come up out of such, requires they see and reach only for the heights of the stars above. If one speaks from the heights, this may inspire others to climb. If we speak from the physical perspective of being next to anyone struggling in the same temporary ways, where are we calling them to come, using words about a fallen common experience. If one need to be inspired, they don't need examples of overcoming fallen physical experiences, being in the midst them. I cannot say I am the same as everyone else. Our calling from GOD is to show us something we do not know, or have. GOD'S new rare experience alone, is what we all need to hear. Leave and forget the old forever, and see only the new. The growth of my spiritual life is not one great wonder, it is many lesser wonders, yet wonders all the same. It is a new way of life to be desired, it is the truth of reality and love to be known and treasured. I say this because I know, desire, and treasure it. Can GOD inspire all? No. It is not a life all will desire. But if a perfect-life sounds like something you desire and would treasure, we hope to inspire *you* to run to become one with what you truly love. My friend, let nothing turn or deter you from the highest conscious truth. Many describe to me what *they* see and hear in Scripture; but they do not hear or see what HOLYSPIRIT shows me. So, I know they do not *desire* to see and hear what I see and hear. They hear Scripture speaking through their temporary physical perspective. This great creation of FATHER has a seemingly magical aspect to it, as it renders to every *rebellious* soul, a false incomplete, barely sufficient, illusion of what they seek. And those who don't want a life under the fullness of GOD, do not know what soon comes upon all living a life without that fullness. GOD is the creative definer of all things, even eternal love, joy, peace, surety, and beauty. What can a rebellious soul expect to find by GOD not being there for and over them. I do not know what they expect, but at their sealing they receive only ever-increasing loneliness, confusion, anguish, and emptiness, all by their own desired choices. They will be provided with nothing they desire, having every endless disappointment of conscious agony, and regret. This is what these should expect to find without end, as this is the horror they have built within themselves; it is *not* an affliction from GOD. It comes only by their never-ending unmet evil desires and unfulfilled evil choices. They have rejected the eternally provided loving ways of GOD, to dwell in the void of having nothing but their forever awareness of the same. If you run from GOD now, don't be surprised when that distance grows forever. Do I try to scare you, well you should definitely be afraid of yourself, as I am. But I more earnestly try to warn, awaken, and enlighten you about eternal love and the eternal perspective. How can I do this without also scaring, because it is scary; and to all who *are* scared, this is good. This fear, indicates some belief towards GOD. We all are filled with both our love and the fear of losing all we love. And to those who read and feel no fear of GOD, I am sorrowful that I have nothing more to offer. You may yet hear words unknown to me, or words I cannot say from someone else in a moment that hopefully will not come too late. Why can words come too late? Because for most, by *their* great internal desires, the great meaning of the one truth of love and reality can only begin to ring true to them, upon their realizing it is too late. Ever so strangely, most souls can and do in pride, envy, and arrogance, hang onto their individual self-rule right up to and past their very last hope of grasping the truth. Can any of these then come to know and express they were completely wrong and hateful? No. These only forever become more self-afflicted and confused. Can these ever truly repent? Can GOD ever accept these? No. At that point it could only be a cry of self-preservation. They are forced to experience they were wrong about GOD and their ways, only after all love and hope are lost. That is nothing akin to desiring and loving GOD above all things while hope and choice were yet alive. GOD has never been in search of those who are forced to accept he is GOD. James 2:19, "*THOU BELIEVEST THAT THERE IS ONE GOD; THOU DOEST WELL: THE DEVILS ALSO BELIEVE, AND TREMBLE.*" Clearly belief in GOD is good, but we read it is not FATHER'S standard, it is eternal love. All angels believe in GOD, yet many chose to rebel. The one omniscient logical patient and loving GOD, takes a singular approach to find *eternal* love alone. GOD searches only for eternal love, which brings eternal meaning, purpose, joy, pleasure, and fulfillment, to all eternal souls, from FATHER above, down to the least. One can never deep down in one's core, truly love what one does not truly desire, believe, and know. To disagree, is to disbelieve. We have all heard it said, I thought I knew them, or I thought I loved them. A great love dies and turns to hate at finding out *what* one loved was not true. If you want to love GOD, get to know them intimately. And you cannot replace a false foundation alone, you must replace the false understanding you built upon it. Like a tall building, the building and its foundation are integral and inseparable. So, a new eternal perspective of understanding, must start from a blank mind, to then hear and see what are GOD'S truth and love. A second chance at life, means a second start in life from GOD. And what if one fails again, do they get a third life, and a fourth, and so on. The KJ Bible says mankind, being born in a pre-fallen weaker estate, gets *one* true and real attempt at a higher stronger rising life. Many think they once made this choice for GOD in the past, when they really did not. Many have flirted with the notion around the periphery for years; but that is not making this true invested firm choice GOD describes. Many may have attended church all their life, and yet, have not gone in true love to intimately find GOD. What GOD means by a second chance, and no third chance, is that if you come to truly know and love GOD, truly receiving HOLYSPIRIT, and learning the one eternal truth, and *then* turn and reject this all again; well, to come back a third time is impossible. So, unless you have known, understood, and loved all GOD'S ways we have written about, and loved his commands above all that exists, having the indwelling HOLYSPIRIT; and you then turn your back and deny your eternal love like Judas did, you might yet be saved. Now if one knows and loves GOD intimately, yet backslides; this is also not the unredeemable denial we speak of. Backsliding is a certain loss of advancing towards GOD, seen in sliding back to a previous weaker position on a certain sin. And like an addict, one can then try again and rejoin the fight. But if one falls too far or waits too long, backsliding can be fatal. In backsliding, the farther and longer you slide back into sinful ways, the less likely you are to recover. Judas back-slid so far and fast, he realized in unimaginable shame, that just like the fallen angels, he had fallen to far too ever get back. Until we are eternally sealed, we can *fall away* into an unredeemable state. Do not see wiggle room, or play a game with this; GOD is never mocked. The stark purpose of gifting free-will choice to all, is for finding love or hate from each. Life can only have its eternal purpose, in our never-ending accountability to our choices. If someone you love dearly in this world, does not care for you in not knowing you; you can still have hope in them. But if they come to truly know and love you, and again, turn against you and lose all their love for you; there is now a difference of accountability and loss of all hope between you. The uncertain ways of all temporary loves can vacillate back and forth, but the surety and certainty of pure eternal love does not. This is where the terms eternal and temporary love enter in. We have all seen those who leave someone and come back again and again. These are unstable souls who cannot know or find what they truly love. They wander from desire to desire in weakness of temporary convictions and intent, without any eternal purpose or perspective. Can any receive these souls back endlessly as if nothing hurtful and unloving always happens. If we remove accountability, and all the higher understanding of what true eternal love means, all laws, rules, and standards of life have no meaning; and all the great agreements of eternal love do not exist. Please see this *Law* of Love, cannot be over-turned or circumvented. It is an immutable aspect of responsibility and accountability that comes upon us all at our introduction into a reality having many free-willed souls, each seeking love, under our eternal GOD seeking eternal love. As each new soul is gifted, all uncertainties rise; yet your soul's accountability to FATHER'S eternal standard remains, as it is *the* highest law of eternal love, even FATHER himself cannot change. If FATHER could or did change this, it would be a foundational change, that must bring the end of all that exists. GOD cannot create love; so, GOD cannot alter love either. Love is always free-willed. One's love is not formed or bound by anything other than one's own forceful desires. Choose to learn only the goodness of GOD, as there we learn, grow, know, and feel the wonders of eternal love. It is an amazing thing that with all the force that exists within GOD, he cannot force his way into your heart. GOD *never* comes into *our* lives, as most believe they do. GOD only comes near to call *us* to come into *their* life, which includes all *their* ways of life. And we cannot go into GOD'S life, without going into their ways of life. GOD only speaks of coming and being one with us, as we come to become one with them. But worldly Preachers teach that GOD wants to be one with all souls in each soul's individual sinful ways of life. There is a divide seen in this, and such imperfect beliefs can only be desired through a temporary physical perspective. Such beliefs make absolutely no eternal spiritual loving sense. Such beliefs are eternally illogical and unsustainable. All such lesser forms of love, fall far short of any eternal responsible faithful desire. Do I claim to have a love for GOD above anyone I do not know? No. I cannot know this. And the fullness of another's, or my love, may as of now, not been found. My responsible faithful love for GOD is yet growing into what it eternally needs to be. And we all must strive to perfect and protect such rarity among us. I have found this eternal higher love does exist, yet I can claim no certain final measure of it, or status in it. I just know and declare the truthful things that exist in me. And these things I know, being shown them by GOD; shown to me *only* because I so desired and longed to see them, and for no reason of self-worth. I am in my fight to find the heights of my eternal love and hope with GOD. I speak of the things in which we all must fight and overcome to reach our eternal purpose. If GOD simply came to us all in fullness, we could have no eternal purpose or work. Life's reason and work is that we see and strive to *fulfill* our eternal purpose; our eternal mandate is to live, and your eternal place is beyond where you now stand. Our created purpose is to find, know, and love GOD above all. This is why FATHER gifted us. I see, hope, and pray, my purpose is forever before me, for if my eternal purpose is always there just beyond who I now am, it forever beckons me into the realm of *eternal* life. And here again, our eternal perspective is required. Do you view *all* things of life through an eternal perspective? Do you discern temporary love, from eternal love? We walk among those holding to temporary love, yet we walk in hope. Our hope is our desire for another's eternal love to overcome their temporary love. Being in eternal love, our hope turns to a pure expectation. We have our firm true belief as we walk through this world, looking to GOD in expectation of their "Holy Promise". Do we wake each morning, knowing this is one of a lesser number of days we have left? No. We know we exist in our never-ending now, within the ever-coming eternal now of GOD. Do we sense a conscious end coming? I sense no end; my soul is eternal. By my eternal perspective in all ways and things, most professed Christians see *me* as delusional, and do not understand me at all. Many think me fanatical, and as having lost touch with all physical worldly perspective. But you see, losing touch with such a temporary physical worldly perspective, is in the definition of finding and attaining our eternal spiritual perspective in GOD. You nor I, can with any *real* outcome, just say wc are a Messianic Jew, or a Judeo Christian; whichever you prefer. For it is only GOD who can proclaim a soul to be a part of them. I do know that in my being born a gentile, I read that I am adopted as a Messianic Jew into the Tribe of Judah through JESUS, who died to adopt and graft me in as one. JESUS being the Lion of the Tribe of Judah. Now I can say and believe I am saved every minute of every day, but until JESUS says it is so, it is just my hope, for I have no saving power. My only power rests in my loving desire to be one with GOD in all ways. To emulate CHRIST in one mind and one accord, in one belief and purpose, and one common walk and way of life. And in aspiring to these many things I speak and write about, I see myself as having no ability to know anything beyond that which I have been shown. That is all I am, I am nothing good beyond my loving oneness in GOD. I aspire to be no more than the loving mind for CHRIST. So, my words and beliefs must be those of the strength and love of GOD alone. They cannot have a taint of mankind's pride or envy. When any ask, who do I think I am to talk and write like *I* am above this world. I say I am taken beyond and above *this* world, and you will hear these words, or not. You will believe these words, or not. I only ask, that you do not see or hear *me*, see and hear only the words. I have no part in the strong reason or purpose of these words. Do not falsely interject me upon them. I am no more than a pencil, or a conduit of pipe or wire; I am nothing. Yet the truth of reality passing through me is a pure mature wisdom. I wish to be invisible, as a perfect window, that light passes through. I sit in no enviable place in this world, for I sit between this world and the next. Hear or don't hear, that is your ability. I can reiterate, or not reiterate, this is my only ability. I am called to speak and write, so I speak and write. I offer nothing more or less than reiterations, and in each word, I matter not. We are each no more, than our work for or against GOD'S Kingdom. Call *me* boastful or a fool without cost of sinning against me; as I already forgive you. But do not disparage these words, for they are not mine. And I know that disparaging them, bears a great cost I have no power to forgive. And if I fail to fulfill my call, or if I add my own words, a greater cost, falls upon me. I know beyond all doubt, that I clearly see GOD'S creation and eternal ways. I know this is my core desire, and my strongest faith resides there. Yet now, at *my* outer edges of learning, like all souls, each formally unknown understanding from GOD, requires I first accept it in blind faith; and this I do. And later its ardent substantiation has always come. If GOD truly shows you something good, you must believe it. Do not immediately demand its deepest fullest meaning; be patient. Yes, we all in growing towards GOD, dwell in the now, where for all new learning and understanding, our blind faith is often required. We each dwell in the same ongoing feelings of inadequacy and hope, as do all who seek GOD. The Holy Angels and the Apostles upon each *new* command or gift, in blind faith and love, accept FATHER'S eternal purpose as being the eternal good of all. HOLYSPIRITS every revelation has never failed to eventually prove itself to me. I rest assured in GOD and grow more at peace, as their every wisdom replaces my own. It is quite strange, that the more I truly deny my own ability and understanding; the stronger, larger, more understanding, and secure I spiritually feel. These are the growing common feelings of becoming one within GOD'S Family. Remember, it is *never* about how high we are in GOD'S order, or who is ahead, or who is behind us; but only our spiritual strength of faithful responsibility, and the direction in which we are going. One does not casually stroll and stop and meander towards their one greatest love; one cannot wait and one runs fervently to it. No, I do not feel that I always run as fast as I should, but can we ever know or think that we do. It is like running towards a growing everything, and running from a diminishing nothing. Will we make it, will we be there? This depends upon our loving desires to be with and of GOD. Will I see you there? I do hope for us all to know and go into those ever-ascending heights together. In finding the one shared eternal perspective, we see conscious projected forces are everywhere in everything. They are in abundance upon every created atom, and in more abundance upon every grain of sand. When we look around, we see FATHER'S endless projections of forcefully entangled physical creations. So, through our eternal perspective, we see ruling creative forces everywhere. And once we always see GOD everywhere in everything, we tend *not* to forget GOD, or sin against them. If you believe in GOD, do you not see that every time you have sinned, you were wrapped up in yourself and forgetting all your loving thoughts for GOD? Do you not see that when deep in loving spiritual prayer and communion, you never sin? Souls run to sin every time they remove GOD far enough from their thoughts. GOD is everywhere, and we cannot hold an eternal perspective without seeing GOD everywhere we look. I know I am an *eternal* soul FATHER has divided out from his vast creative greatness, and set me free as a willful creative awareness that can never cease to be. But though we eternally own and are our soul of creative force, we are still completely restricted and dependent in our connections to the *required* physical half of reality. And in our dependance, our soul can be bound to a body that is an immobilized unfeeling prison if determined by GOD. Our soul can be aware of much information, yet not be able to strongly feel or respond in any desired satisfying way. If one's body's use of certain forms of information is weak and stochastic, that soul cannot find love, peace, or pleasure; just anguish. All lost souls are in the early stages of this now and don't recognize it. The sciences think they can see pure randomness, but to perceive anything, requires a patterned complexity. FATHER sees his aether in its pattern. The fallen can't perceive the complex boundaries of reality, nor comprehend how limited their fallen sensing abilities have become. They think they can physically sense and know enough to make certain claims about reality, yet they can't even see their grave errors. Yes, the cause of *this* paradox is their temporary perspective, being an inverted argument, full of physical miracles and false faith. Sciences knows that even with their enhanced technology, they are still greatly limited in the 6% they now physically sense; but their greatest limitation resides *in* their desired perspective and belief. They cannot accept that there are far lesser and greater aspects to reality beyond their fallen desires. And in abandoning all belief in an aether and the soul, because they cannot measure them, they will never make any true sense of reality. Seeing a pure Godly science answers every question about reality, why deny it just because you cannot physically sense or test it; that is insane. The physical senses can never give you anywhere near the whole or true story, and you will only find more insanity. No consensus of physical belief, formed through a temporary perspective, can ever know anything final and firm. The required medium for reality, is

FATHER'S Aether; being his forced, patterned, malleable, extremely porous, 3D, spherical created structure. Once you say that only in seeing, is there believing, you are trapped in a limited physical belief, as you can't physically see very much. It is only in faith that there is far more, that we can see far more. Our soul's senses joining the higher consciousness of GOD, can comprehend the Aether. But science, without a scintilla of proof, denies this aether exists. Materialist are like babes with awkward hands, curiously fumbling to bring a toy reality close to touch their face and mouth to physically sense it. GOD created human bodies in a strong amazing sensing fashion; but the physical is the far lesser part of who we are gifted and created to know. When souls are falling, they see only the physical, and cannot spiritually sense, study, learn, and know the greater ethereal things. They fantasize they are explorers on top of the world. But they are on the bottom; and being there, they still look and dig in the wrong direction. These have built a desired allusion of self and reality in the mirror. How simple an equation it is to deny the truth, to see the physical half incompletely and obscurely appear to work on its own in a godless temporary perspective. Materialism stands upon faith in their standard model, which more, and more, is seen as insufficient, and it is built upon a physical miracle. But they must find a physical replacement before letting go of the broken. In their ever-growing doubts, they hope they may still be right. All temporary perspectives are no more than loose malleable belief structures that allow souls in this world to forget their eternal accountability to GOD and sin. This will not work in the hereafter once they find they are an eternal soul; and all too late, glimpse the eternal perspective. We see hope for a few scientists who don't see materialism as proven, and have not yet denied their soul. We see the term *super*-natural; but "super" means a greatest highest measure. So, a super-*natural* GOD is not an un-*natural* belief. GOD is the greatest highest *natural* form and measure of belief. We must also see *forgiveness* through the eternal perspective. So, what is true forgiveness, and who can truly forgive? Scripture says there can be no eternal forgiveness for betraying GOD and all creation, without sacrifice, suffering, and pain. Does this sound like vengeance? Well, it is. Call it retribution, penalty, or justice; these terms are all about accountability, responsibility, and recompence. All things of value must be re-balanced and made whole. We read that the pure and truly *worthy* can choose to suffer for another's sins; and that FATHER cannot refuse acts of eternal love. The imperfect can verbally forgive each other, and the debt between *them* is gone; yet that does not settle their debt with GOD. Only MESSIAH'S suffering, can pay for the *past sins* of all who truly *repent* and renew their ways. By this, and emulating our loving example of CHRIST, our own great strength can be found. Eternal love blooms from a *pure* desire, and wins against all odds. Eternal love overcomes every obstacle in its strength and will. Again, how is love the strongest ruling force in all reality? Simply put, love is the strongest ruling two-edged desire within GOD and us all. Materialist say love is just a chemical reaction to help perpetuate life, and all complexity appeared by chance. Yet in their next conversation, they admit conscious reason and purpose do exist. Is this a contradiction, or semantics? Why is love so painful and costly? By loves great pain, we all know it is a spiritual desire and awareness that transcends all boundaries of every known physical correlation. There is not one empirical evidence that says any physical aspect or thing, can in any way or form of complexity, know or be *aware* of its own existence. Yet we are aware there *are* the purely physical aspects of creation, and the unseen founding aether. Pure physical non-complexity on its own, minus projected patterned force, cannot align or orchestrate a single thing. All evidence says a pure physical singularity, cannot generate a forceful fluctuation on its own. But countless pure physical singularities can be aligned by an exterior quantized patterned encapsulating force, that can be stored, radiated, restored, or diminished. But force can never originate from a physical aspect. New force originates only by every soul's intentional desire to exist; being our conscious awareness of life that creates particle-wave based perceptions. And when studying the generated projected forceful aspects of our desires, called love. We see an ongoing generated projected conscious force of desired intent, purpose, and reason, that existed in no prior moment or space before that love was born. And we see all aspects of *eternal* love are directed at perpetuating all things together in caring, sharing, sustaining, and preserving *every* good spiritual and physical complexity in each other without end. We also see that just as our eternal love for GOD requires our sustaining forceful aspects of nurturing and cherishing; that hate for GOD must also be fed. In other words, all desired emotions must be temporarily *or* eternally fed. So, we see that if our eternal love fades, it was *not* fed. So, pride, envy, and lust, can also be starved out of existence in our moment of choosing. Please feed all good emotions, and starve the bad. This is easier, once we see the need for love to be returned to FATHER, was once so great, being utterly alone as the only conscious entity to exist. In knowing this, we see the truest greatest value of eternal love is known only to FATHER, who experienced a loneliness only he can know. Such a lonely want and need for others, sounds unbearable. FATHER so loved his SON at his gifted creation; before any other soul had ever existed or loved another. And FATHER was loved back by his only begotten SON, before any other soul had ever experienced or returned loved. This *is* the first strongest love to ever exist between souls. FATHER'S experience of such loneliness, is an eternal lesson we must understand now, not needing to experience it for our self; or we must and will learn it the hard way forever. If GOD'S first love was his strongest, does love then come in lesser orders with more and more souls. Is FATHER Holy, Righteous, and Just, in loving JESUS more than he loves us? Of course he is, just as we love JESUS more than all other people and angels. Do you see how and why, in this lineage, there must be that last omniscience of the weakest form of eternal love for GOD. Which must appear as a divide, just before the first weakest form of eternal confused hate for GOD appears. Being the strongest temporary love for GOD that fell just short. So, the final number of souls in Heaven will be set by the standard of eternal love for GOD. And might it be an eternal perfect equation. We must seek the immutable understanding of this; as this too is an aspect of the eternal perspective. I hope you clearly see that all *true* things, emotions, understandings, and beliefs, can only be held through GOD'S shared eternal perspective. And holding an individual perspective, tainted in all its timely temporary ways, separates one in unsustainable weakness from GOD. Do you see the opposing directions of force between the temporary false perspective of literally everything, and the true eternal perspective of the same. I see everything in one truth and shared understanding, yet I still have much to learn. I see how, not just the world itself, but more so the souls I love in this world, slow and hold me back. I know I cannot help them all, and any temporary love given any advantage, can bring me or anyone down. I am of no value or help to anyone, being with them in a temporary individual state. Most often, our family and friends who are still individuals, can only be helped by others, as most reject the wisdom of those close. We see engrained fallen worldly familial and societal traits, called sibling, and associative rivalry; also, generational rebellion is real and growing stronger. This is a sad, true, and destructive fallen aspect. I have so often wished GOD could put a switch in front of me that I could flip, to sever all weaknesses within me. But the immutable reality of free-will, desired, chosen, and learned, conscious existence, make that impossible. It is all about our learned growth and direction of all streaming choices coming to and through our desires; and our loving reasons and purpose we ascribe to them. This is life, this is what our conscious life of desire is. This is the core of all conscious reality, and the true *enduring* spiritual pleasures derived from life are only found through one path and direction. In conscious death (conscious separation from GOD) a soul feels only a temporary purpose and understanding of temporary physical love, void of true spiritual pleasure, while not understanding why. They sense endless growing confusion as to where, and why they exist. For such souls, all things forever diminish in meaning, as they empty of all love, being self-aware, sensing little to no physical pleasure. Am I preaching fire and brimstone? Yes. For such souls will be eternally sealed in indestructible bodies that feel little, set under fierce gravity within a fiery prison created by FATHER, known as the lake of fire. We only sin by deep desires for things and ways of life, FATHER has not made for, or given to us. If we cannot overcome such desires, FATHER cannot allow us into his final Kingdom, where such desires would cause it to fall again. This choice process is about finding souls who, having seen things and ways not made for us, truly do not desire such things. This lake will hold all evil souls, memories, and all separations of death and hell within. And it will protect all souls who dwell just beyond this lake, within the gates of Heaven. These lost imprisoned souls may sense uncomfortably sweaty heat, but no searing pain from the fire. But according to their sins and desires, they *will* feel their own level of eternal conscious torment, being dry and thirsty in their unmet unquenchable desires. Surely not a place to want to be forever. But that is the eternal reality of all *individuals* within it. The GOD they did not want, they will never have. These *individuals,* wanted to throw off GOD'S good and loving created ways and provision, and be a god over themselves. But when you cast off GOD'S created ways of eternal provision, *all* that is left is your awareness of unanswerable questions and unmet desires. These will know how undesirable and small they are. The term "prison" is defined as a secured harsh undesirable existence, meant for contemplation of a soul's destructive ways. Prison is to deny souls all pleasant things, and keep them from harming the good. This prison must be sustained by the eternal conscious efforts of FATHER, and I am sure he is pained at his great eternal loss, and work to detain so many souls. Let's examine our search for GOD from another angle. If we seek to truly know GOD, we begin a journey of finding the many things we *want* to know and believe about GOD, and all we did not expect, yet *need* to know. Searching for GOD from an *eternal* perspective without a sufficient love for GOD will fail at some point. But searching from a temporary perspective is *always* fatal. Most souls searching for GOD, take a wrong first step, in a wrong direction, by searching for *a* god of their own desires and design, and from a temporary perspective; but do not see or admit what they do. A true search is not about finding out what I or you or anyone else wants to believe about GOD, it is about finding out what FATHER believes about himself and us. We know all do not go to Heaven. So, it is learning where FATHER'S own lines of created eternal life are drawn, which is learned in our spiritual fine tuning and growth. Souls who search *from* a temporary physical perspective, rarely consider what FATHER believes. Many think FATHERS beliefs are like theirs. I immediately feel something so terrible, when anyone says to me, you have your beliefs, and I have mine. I reply, I put aside all my beliefs to search for and adopt FATHER'S beliefs, but they cannot hear this. They just say again, but that is your belief. It is frustrating that most of what we reiterate from GOD, cannot be heard by most. This too, is the divide between the eternal perspective, and temporary perspective. An eternal perspective, is a Godly perspective. An eternal perspective knows by being *under* GOD, our understanding, born of only true questions and answers, grows endlessly, yet never changes. While individuals desire their ever-changing questions, answers, understandings, and beliefs. Deep down, individuals do not desire to be eternal or under GOD to any strong degree. If you chose to be under your own guidance and provision forever, you must forever diminish yet forever exist. Individuals tie their entire happiness to the body they now use; a as the body fades, so does their joy. Individuals are trapped in a revolving door, not desiring either of the only two exits. Having a wrong perspective is like being in a bad dream that never makes complete sense and never ends, yet gets worse and worse forever. I can attest, that if you truly and fully receive and adopt GOD'S eternal perspective, it feels like waking from that bad dream to an amazing calm bright reality. It has a true real purposed feel to it, that grows. I love FATHER for countless reasons, and one, is waking me from my temporary perspective. How do most think I am not happy in the way GOD shows me eternal life. They see with blurred eyes and perspectives that see GOD in personalized ambiguous ways. I hope my excessive rumination does not bore or offend. But I know the more ways, we see and hear a certain thing, we attain a firmer understanding. Of course, it is still the good core desires within us that allow this matter to implant and take root. It is only by the strength of a soul wanting to hear, that another's speaking becomes profitable. Do I seek to move and pull all to GOD? Do I seek to find every way to open and expand everyone's eternal desires and perspective? Do I seek to change every mind that does not agree? Do I proselytize towards each and all to propagate a strong Godly bias within their self? Yes, as I am commanded; and unapologetically so. In these days of political correctness and cancellation from public life, I am an easy target in this world. But what haters do not understand, is that I cancelled myself from this world long ago. They cannot tell me they are kicking me out of their hateful club, to which I in no way do, or desire to belong. They declare they have the right to kick me out of this world, and it infuriates them, that I care not. They can have this fallen dissolving illusion, for I seek the eternal. They may even kill this body that holds me here, but they only free me to go home. It is my great reward to finally go to GOD, and I long for the next world; it is they who need to fear death. If I have done all the good for GOD'S

Kingdom in this fallen world that I can, I have no reason, purpose, or desire to be here. I abhor the perverse ways of this world. I look around and see only lies upon lies, deceit upon deceit, destruction upon destruction, hell upon hell, and death upon death, as the stench of it grows more and more. Many desire the temporary surface facade of this place; I do not. What do you see, do you see a world based upon and filled with physical beauty, love, and life. If so, you see only its temporary fading surface *illusion*. Please look beyond this façade, and see where all dark realities come from, and where they all go. I see a corrupt fallen, yet once perfectly designed physical foundation and structure. I see a once perfect hope, now corrupted and defiled. I see truth now twisted beyond recognition. I see the growing emptiness of nothing, as it is all going nowhere. It is all hollow and poured out, and has nearly run its diminishing course. See this world compared to GOD'S perfect Kingdom, and you can see the lie of this world. We all soon leave this vanishing world of choosing our belief and understanding. And that is all we will take from it, being our choices of belief and understanding of love. It is these things that declare to GOD, who and what we each desire and chose to be. With all my eternal hope and desire, I want to be one with GOD'S truth, goodness, love, peace, and joy; being the most beautiful and desirous reality that can exist. Once you see this clearly, it is only this you want. Who can gaze upon and understand such a wondrous beautiful thing, and not desire it above all. It is the greatest-conscious-qualia of all our highest created emotions. If you could float outside of all GOD'S created realities to see the chosen directions in all souls, from where FATHER watches, do you think you may understand this all better. But to see this truth properly, require HOLYSPIRIT and our selfless objectivity. *And* if we are totally selfless and objective here and now, and have HOLYSPIRIT, we need not step outside of reality, as our view of reality from the inside or outside will be the same. To adopt FATHER'S view of all ongoing spiritual separation, requires only that we replace all our subjective fallen ways and beliefs, and we need not leave the physical reality itself. GOD'S perspective changes not with from where, or what, one focuses upon. *All* that physically exists, is relative and entangled by an eternal perspective. Now on a far smaller inferior scale, we are in the essence of our soul's creative forces, like FATHER, we are apportioned out from what he is, creatively. We are all the offspring of FATHER at our soul's gifting. And our free-will allows us to become one with GOD, or reject them for individual desires, beliefs, and understanding; to want to be one with GOD, or to be an *individual* against GOD'S creation; to be fulfilled in being all we are created to ever be, or trade it all, and become the most hateful anguished confused soul, you could ever be. Would you trade eternal happiness for the eternal ownership of one grain of sand? Yet, to eternally own the quantum properties of one grain of sand, being imprisoned on the lost side of all final separation, would make you the greatest, least suffering soul there. Who can know all of this, yet still choose to be an individual. How can we get an individual to see the one pure truth they so desperately do not desire to see. When summed up, our free-will is a choice between eternal loving joy, and self-destruction. And paradoxically, the few choose the first, and most choose the latter. And those who choose the latter, in self-preservation mode, declare a foul has been committed by the few. They can't accept that after choosing individuality, they can never be free to eternally spread their own chosen ways of life. They see this unfair, and the game as being fixed. These cannot see their great eternal flaw is in their own choices against all that is true and eternal. But we see that all that is eternal, was true long before we ever existed; most act as if reality began when they did. By free-will choice, a soul's core desires can be wild, chanceful, and not contemplated or understood; they choose most things by emotion. Their every desire forms its own evil logic, its own foundation, and conclusion. We see how our core desires, determine our beliefs; and all individual beliefs are of their own desired insane logic. A conscious *law* arises, being that every individual's core belief and love, are not eternally compatible with any other individual's core belief of love. Now in souls who question free-will, a need arises to prove they have free-will. Rebellion is strong-evidence of free-will, sensed on a soul's surface. So, many opt for rebellion, even in it being against their every obvious eternal interest. But for me in seeking to live GOD'S perfect way of life, I sense my free-will in its greatest measure. I am completely secure and assured of my own free-will, as I *must* desire and make every faithful responsible choice to love and live this way eternally. The children of GOD, strongly exercise our free-will at every step to perpetuate a common loving peaceful life in every next free-will choice before us. While those *sealed* as lost, have made their last choice of existential consequence and true meaning. It is only by casting away all shared eternal responsibilities, and choosing individual ways of unbridled change, that souls lose their recognition of free-will. Every individual understanding is born of physical and conscious confused disorder. It is by GOD'S eternal existence, that only two logical directions exist: one towards GOD, and one away from GOD. And in the direction away from GOD, there are endless *individual* illogical paths; but only one logical shared path towards GOD. The basic core logic of many souls going in one direction upon one path to GOD, will be one and true. Eternal love has an extreme purposed consuming desire to be needed and purposed. In eternal loves founding and growing, it has its learned and calculated understanding to protect and nurture all good enigmatic forceful-phenomena. But in this world we see love appear, but in not being properly aligned or protected, we see love most often fades and dies, making it temporary. Love is life's only purpose, so we all want and search for love. We truly live and exist for no other purpose than love. But can we individually know eternal love; or only in one mindful belief, working together. Love is our ruling forceful desire to exist. Who can describe love, beyond its being the highest of all desired conscious qualia of purpose, reason, and joy. The strangest aspects of qualia, rest in their being experiential and quite indescribable. So why is love such a differently described desire among individuals. We immediately sense love, and try desperately to describe it. It is always that which we desire most, but most souls feel, choose, and love the physical side of reality more than the spiritual. Any physically founded love is doomed to fade and die just as everything physical now does. Only the fewer, feel and choose the loving spiritual eternal side. I say without fear of blasphemy, that the most basic forceful essence of every eternal or temporary love is the same; but what it is projected *at* by most souls, is many contrary things in many differing strengths. Love is not ruled over by GOD, but love in the eternal sense, rules over GOD. It is who and what GOD truly is. Scriptures declare fully and in every way that GOD is loving. GOD is the highest purest most forceful source of love that can ever be. If someone would ask me to describe the mind of GOD in one word, it would be loving. If asked to describe the soul and spirit of GOD in one word, it is loving. If asked to describe the need, the actions, the ways, the thoughts and hopes, the inside and outside, the comings and goings of GOD in one word, it is loving. Why is love in the creation? Why is love in this world? Why is it everywhere in its endless attracting creative and contrary forms, always there with its great pains and losses? Why do souls desire to exist? Because we sense and seek this quale called love. Why do people who have no love, want to cease to exist? If a soul loses their every loving desire, they lose all will to live, and they die. How is love the force of wanting to exist within us all? It is one of the true universal constants of reality. Why does the desire to exist fade when souls choose to love the temporary things of existence from a temporary perspective. Because life takes away all those temporary physical things, which destroys a soul's faith in physical love. Zero eternal love, brings zero faith in life, and no desire to physically live. Such souls must and will forever die more and more. Yet as long as these souls sense enough physical information, even if it is in confused anguish and misery, they must and will forever exist in their awareness of it, as they desire to cease to exist. This is a never-ending conscious state of confusion and separation GOD calls the individual abyss of hell. In its young and early stages, while still in this world feeding on temporary physical love, its intermittent losses are endured. And to new young life going by so slowly, eternity seems a bit much. But after just forty to sixty years, a soul realizes life is always just the current moment, and it fades in the blink of an eye. The source and provision of eternal love, is waiting and wanting to secure us without end. And so, we see why *our* greatest loving obedience must be for the source of it all. Seeing and understanding reality through the eternal perspective, is vital to nurture and protect the good and shun all evil. Our core love must be true, it being our strength to hold to GOD, who cannot hold onto a dying hand that has let go of them. We all sense by our experience, that eternal love cannot work in a dying temporary way. If two loving souls hold onto each other, then one feels the other not love and let go, the heart that still loves breaks, and the grip of one hand is painful as the other hand loosens, pulls, and slips away. There is an immutable reciprocal law of love. Thus, only love can feed love, love needs love, love wants love, love loves love. Just these four letters (in the English Language) form the biggest word that can ever exist. And these glories and agonies and choices of love fill GOD and all souls, and all we perceive. Some ask is perspective the ruler over love? No, love is the ruler over perspective, love is the pinnacle, acro, and peak of all conscious reasoned purposed reality. Love is the strongest above, and flows down through all things. You cannot describe one conscious thing in all reality that is utterly void of at least love's outer fringe aspects called hope, chance, and fear. Do you see this is what all chance, hope, and fear are, at their core. Such feelings are our first ethereal aspect in our search for the reality of love. And if only a temporary physical love is found, then every meaningless physical chance occurrence has power and influence over that love. Physical chance in this realm, is about finding and then losing physical love alone. A physical belief alone, finds only fading attractions, and never any eternal desire to build protect and preserve love forever. We see in every eternal consideration, that *physical chance attraction* has no part in its reasoning, purpose, or conception. The strongest *physical* information projected between souls, can affect them in powerful and evil ways; yet our spiritual choice is where we resist all the ways of this covetous world. If one desires the physical most often, they must feel and exist in a temporary physical perspective. One who desires the spiritual ways and things, and denies the physical, has the spiritual eternal perspective. And we should quite easily be able to see which we are building. I try often to sense and equate myself in this regard. During the early mornings when I write, I feel mostly spiritual; but sometimes by afternoon, I feel like I am barely hanging on at 51%. I have much room to grow, and much I want to do for GOD. I feel quite spiritual during my long morning talks with GOD upon waking. After which I get up between four to six a.m. to write; these are my strongest hours. So, I know my eternal path is in learning, rebuilding, growing, and loving. And at a certain point we begin to better feel and see where we always are, and where we yet need to be. For me, this is the only path to be on to understand myself and my essence of love, and how I relate in these to GOD. When we all look out into the vastness of space, what do we see? Do we see only the physical aspects of reality being there and moving by its self? Or do we see FATHER has *perceived* his firm vast creation, for us to also see? Do we realize our lesser forceful abilities FATHER gifted to us, so we can also perceive? At first it can be hard to become one with GOD; but with all gained understanding and growth of love it becomes easy and pleasurable. But no matter how hard, easy, or pleasurable it becomes, it is an eternal step-by-step learning growing process. Is it harder to believe that we as souls control the body, and create particle-wave-based-perceptions of GOD'S Creation around us? Or, is it harder to believe that all unknowing, unaware things, just move and create themselves without any reason or purpose; and all our complexities of perception and awareness just came by the unimaginable odds of physical chance. It is impossible for *me* to believe GOD is *not* there, and that all reality and life and awareness just happened by such impossible odds of chance. What *I* see is that which knows, moves, creates, and rules over that which knows nothing; and with zero evidence of cause, science says that which knows nothing, creates and rules over, all that knows. I see a great act of supposition and false belief in all individuals who create their perceptions of GOD'S vast external reality; yet by desire, see it, as if it has been there on its own and of itself for billions of years. Now we inherently know we each are not *great* Gods capable of creating the universe. Yet we know we had our new lesser conscious gift, as our soul began to form our understanding through our created perceptions. We also know that what we outwardly commonly sense, preceded us; which is a proper and true view. But if you do not see that GOD existed before this reality, you must accept endless physical miracles that overcome incalculable impossible physical odds; and that all occurs for enigmatic mystical physical reasons. And top it all off with the far greater miracle of a non-physical entangled quantum illusion of awareness, appearing out of nowhere, for no knowable reason. You must also believe these physical miracles waited eons before suddenly emerging by chance occurrence; far longer than FATHER'S six-days of forceful entangling the complex layers of his created firm perceptions into existence. I firmly know that even the simplest thing of one plus one equals two, could never exist without a soul to contemplate, define, and know it; and every affect and effect was created by a soul. All logic and complexity exist at being forcefully perceived into creation; this is the truth of the eternal perspective. A temporary perspective does not see far, broad, or deep, enough to know reality. This is a dangerous weakness, as both love and understanding of GOD are our only true strength. Why do many view the forceful created provision of GOD through such physical restraints. We all see we cannot consciously create our external provision, yet most, looking past GOD, see a right to see and use it in their individual ways of desire. They turned a once perfect eternal created provision into an imperfect temporary provision. They think they take hold of this world as theirs. Most believers can't see a *loving GOD,* being anything other than what they want him to be. These believe if GOD loves them, he will acquiesce to their wants. These believe GOD'S love exists in giving them what makes them temporarily happy; even if it sinks their souls into the confusions of hell. A temporary perspective cannot see far enough ahead to see what comes by evil desires. But *our* perspective, sees what comes from all true or false perspectives. We need the eternal perspective of GOD to be consciously held firmly, in order to walk, understand, live, and love, without end. Or our soul's understanding must forever drift away from GOD. In our early awareness, most don't remember, we all heard and saw things from GOD which took root, or were rejected. These came not from any Church or person, and my family rarely espoused anything of GOD, or *any* true understanding. Now I am also old enough to remember in 1959' 60' 61' the daily morning K-2 class prayer, held in all public schools. Then godless materialist sued, and a godless Supreme Court banned prayer in school. GOD was grieved and stores up much wrath from that ruling, as great hell and destruction followed. I would not want to stand before GOD as a supporter of this, or a godless right to murder unborn babies, or restrict religious expression, or legalize ungodly marriage, or promote racial or sexual divisions, or divide the land of Israel against FATHER'S borders; or any evil politics that are against GOD'S commands, statutes, and ways. Since the fall of mankind, liberals have worked to destroy GOD'S laws, and interject division, destruction, hell, and death into the lives of billions of souls around the world. And judgment awaits all who continue to help and follow in this. How evil can this world get, as the insane masses cry out for more and more? I see only madness that yields no eternal reason or purpose, as it feeds their appetites for evermore insanity. According to Prophecy, this world will increasingly lose sight of truth and reality for a few decades more. Again, just how insane can this world get? Please know this truth; certain depths of insanity do not allow one to return back to the true understanding. The world, en-masse has passed that point of return. The world at large will never wake to say, ok GOD, we have had enough insanity. No, they will want more insane things; and upon getting those, want more again. This world is set on fire with insanity, and those burning do not know it, yet revel in it. It cannot be stopped by anything other than FATHER cleansing this world of it in the Tribulation, before CHRIST'S soon return. FATHER'S cleansing is the same as with that prior great flood in removing all insane souls from the world. Not with a flood this time; but first by mankind using biological and nuclear weapons, and engineered famine and disease. After these, two hundred million Holy Angels cleanse it further; and only by the mercy of GOD, a small remnant will be left alive. Then the Millennial Reign of the last souls, choosing their final belief. There are in this world, many forces of partial and lesser love for GOD. There are in this world the fewer forces of a strong love for GOD, and fewer still, the forces of eternal love for GOD. Billions of weak confused souls, fill churches, synagogues, temples, mosques, and what have you. We need a physical refuge, prepared by GOD to gather *us* together as one in CHRIST. If any believe this is an impossibility, they have by such belief, made it impossible for themselves. I know it is certain, for GOD has said it will be; and it will be for those who love GOD above all. All who love FATHER'S ways, are the far fewer, when we should be the far more; *we* should be all souls. But through it all, as our desire for GOD to rule over us, and our eternal perspectives grow, we see evidence that we are finding the one true GOD, and are being found by them. To become one with GOD is an other-worldly religious experience; it is finding and growing in the one shared Holy of Holies belief and understanding. This is our truth and way of becoming one with GOD. It comes not in saying or reading certain words, or participating in church functions or events. Such things are ordained to be good; but our becoming one, rests in obeying *every* word of GOD. And understanding and experiencing our existential purpose in GOD'S created ways. How much do we deep down in our core want to truly and intimately see, *and* be seen by GOD. Do you wish GOD had created one, or a few eternal things differently? Then you must learn and know why you are wrong about those desires. Imagine if GOD brought a million souls into Heaven, and each held only one contrary desire, in how to live eternally. Now we know a million souls will not hold the same one contrary desire. So, that would allow many contrary divisive group desires to start dividing Heaven. My core desire is only to find *eternal* love, peace, joy, and life for all; nothing else can matter. Once we understand and accept that it is all about one common faithful responsibility, and not about endless physical desires that divide and disrupt, we can see through the eternal perspective. Do you now, in this world, project and protect spiritual life and love above all physical desires in each moment? Do you see eating as a desired experience, or a physical requirement? Do you see physical congress is a marital procreational act only, or do you see a recreational experience of physical pleasure? As all physical experiences are desired and embraced in godless ways, all Godly ways are lost. Physical pleasure has no eternal value, remember they are only temporary patterned information running through our required physical platform, that allow us to relate to one another. The physical semblance of the world, from its perfect beginning, to its fall, to its current state, and on to its dissolution, matters not. Focus upon and remember the commands and ways of GOD that remain through it all in every moment. Only our eternal loving reasons and purpose go with us into the next life. The created ways of sharing life through physical information, are not meant to be about the physical side of all information; but about the spiritual love and peace our platform enables us to relate and sense. If our physical environment is our ruling focus, even for a beautiful sunset, our eternal perspective is not in that moment. We must see all physical beauty as the lesser aspect, seeing GOD, the creator of each sunset's beauty. We cannot help but know a pleasing sense of something; yet we can know it through GOD, or physical desire. Keep all physical experience under the spiritual experience of the eternal flow of love, and your soul will reign over that experience. Many don't care about losing the Spiritual experience of life, as they seek to fall as deeply as possible into the next and next pleasing physical experience. FATHER could eternally correct every physical error in the patterns of life, caused by every self-centered experience. But we must understand that FATHER has no desire to do so; he has not searched for an abusive family to forever clean up after, who only use and abuse his loving provision. FATHER was *first*, and *then* all souls were gifted and invited to spiritually experience the eternal good created things of life with GOD and each other. I do not want GOD to *ever* have to clean up, or repair what I have done, for it shames me; let alone do so in perpetuity. Each day I strive in my love to be considerate of GOD'S desires, needs, and created ways. I want to please GOD, because I do truly love GOD above all. To me, GOD is the most beautiful desirable personage that can ever be. This is not hard to comprehend; every aspect of the eternal perspective, are of a most-simple perfect logic and love. What is required, is for us to desire only eternal things; and through a growth process, all these things align naturally. When we hold our *most* loving gaze upon GOD, their love flows through us. But if we turn to love others more than GOD, those who cannot eternally hold us, we have broken our eternal bond to life. The reality created by GOD is more beautiful, meaningful, and loving, than this world can know; as it was pure, perfect, and spiritual. If you study the spiritual and physical aspects of GOD'S coming eternal Kingdom and New-Heaven, you see every aspect is perfect, loving, faithful, responsible, and logical. Looking at the physical aspects of Heaven, we see every provision of Heaven is created through FATHER'S design and labors to form it out from his aether's points, that are still his body. These designs must be forcefully enforced by all, to be eternal. We cannot treat these in a temporary fashion, as if GOD forever fixes all things anew each moment. And in our being only a soul, we can't sense, experience, or move about, and share our spiritual realities without a body designed to freely and richly sense the creations of GOD. These current bodies are far fallen, and do not sense near what they once did. Yet they are all we currently have, and we can still learn all we need to know by HOLYSPIRIT. We see eternal care and labor are required on our part to maintain our physical space within GOD'S great reality, we must clean, repair, and replace things. If we have children, through efforts of love, we provide many things of spiritual and physical life. And if we experience, or witness, a rebellious child, we understand something else. We know that providing for a loving grateful child who cares for, conserves, and preserves the things we provide, is rewarding. But endless giving to a rebellious destructive unloving children who waste, loses, and destroys all you provide, will destroy your love. Now add eternity, and hundreds of millions of children into this equation and you see that FATHER cannot work to repair all destroyed things for eternity, there being no responsible faithful love in return. There must be a symbiotic mutual love between a provider and receiver. Any who believe FATHER can responsibly, faithfully, and eternally love a soul who does not responsibly, faithfully, and eternally return that love, has misinterpreted Scripture. FATHER loves those who love him, and gives his physical provision; all we can give back is love and peace.

We are far greater benefactors in the loving FATHER Child relationship he seeks. We see no unreasoned un-purposed unloving choice on the part of FATHER in all his eternal ways and creations. FATHER'S created ways, are the ways he omnisciently knows he can lovingly sacrificially give and do all this for us, for all eternity; and on balance, teaches us what he seeks and needs in return. In other words, like everyone, FATHER has his loving limits to what he can do for each soul in Heaven for eternity. If FATHER had no limits, there would be no reasoned separation of souls at the end, would there. But we all strongly sense a coming growing eternal separation. GOD'S labor, effort, and works of love are like ours; they are choices made to do things out of love, and needing to be loved back. Please understand, that when FATHER said we are all made in his image. That means his physical creative *and* emotional image, his needful loving image, his hopeful image, and his forceful projecting image. GOD has their desires and hopes to love and be loved. And we, like the complete image of GOD, have no desire to be used, taken advantage of, abused, or to forever provide for others while being unloved. Have no doubt that each selfish soul who is irresponsible and unfaithful to GOD, is a burden to them of eternal emotional hurt. There is no righteousness or holiness in GOD being taken advantage of and abused and unloved for all of eternity. GOD mercifully endures the over-all, and individual rebellion until the last love is found, and all further hope is gone. Yet GOD finds that most go all the way to the grave, without ever considering their eternal love. Most can't accept surrendering their individual choice to define for their self, what is good and evil in their life. GOD in an omniscient place, has drawn a good and perfect line, that rests between the free agency and spirit of each soul, and the oneness of an eternal loving family. Their line defines what it means to both, be ourself, and yet one with GOD'S Family. The line where we each being one soul stops, and where we all as one eternal family begin. FATHER, SON, and HOLYSPRIT, are one in this way. The Triune GOD and the Holy Angels and Earth-born Saints are one family in this way. GOD teaches and shows *us*, and *we* see his creations for what they are meant to be. These are not arbitrary unreasoned non-purposeful lines, they are Dualisms actual boundaries of shared eternal love, born of GOD'S perfect understanding. Boundaries that when crossed, without fail, cause love to diminish at least in a perceptible way to GOD. Just because in our lesser form, we cannot sense any destructive problem in our desires, does not mean GOD does not sense problems.

We trust GOD, because if they did not love, love could not exist. There is incorporated in all the good that GOD creates, their perfect loving understanding and agreement of what it is to cherish protect and sustain a loving family for eternity. FATHER in omniscience, applies this standard to the coming New Kingdom. So, it will be designed in perfect accordance with all eternal needs and provisions that must be. Do you know what it is like to try to teach someone knowledge and truth that will guide, enhance, and make their life eternal and loving? But they just can't hear it or see it, because they don't want it. This is how GOD feels, about their omniscience. And this is how *we* now also feel, holding our current level and share of omniscience. Yes, our iteration of GOD'S truth is our part of GOD'S great and full omniscience. If GOD offers their omniscience, and we receive it, we become omniscient in our levels according to all we receive. Do not think it evil to say that when GOD gives you their truth, and you fully believe it, you are then in that part become omniscient with them. Omniscience is the joining of "omni" meaning all, and "science" to become a compound word that means knowing all science. GOD offers all the knowledge we need to be omniscient and eternal in love and life; but most can't see this, and can't wrap their minds desires around this truth.

Now please see how your growing level of omniscience is a growing rising part of the eternal perspective and mind of CHRIST we are all called to seek and find. Our eternal perspective brings our oneness with GOD in many rising ways; and please do not limit yourself or others, but let GOD set all limits as to what you or I can, and are meant to be. Now we see and know that FATHER receives zero returned love from the physical side of reality. Our gifted free-will souls are the only source that can satisfy FATHER'S loving needs. There are the lesser *spiritual* loves of each saved soul, and there is the greater *spiritual* love of oneness among all saved souls. Please see that eternal love resides in all the perfect shared aspects that wait just across the line of where the part and facts of you and I end, and where the part and facts of us, with and in GOD begin. Eternal love is a true eternal bond of oneness, a togetherness of understanding and belief as to all eternal existential aspects. Please learn of and love GOD and family, and don't choose to be a contrary burden and abusive destroyer of GOD'S loving created gifts towards us all. In its essence, rebellion is the rejection of the created ways and realities of existence. And as all realities were created and defined long before we newly came to be; this is now a *devolving* reality, as each new generation rebels against the former. Adam and Eve rebelled against GOD, and Cain rebelled against everyone; so, in just two generations, the first murder appeared. Can we see how far mankind has fallen and devolved from GOD since then. Surely you sense that we write about these things in many ways, from many vantage points? Does this bore you, or do you in interest, sense their entangled expanding aspects? Do you see how one eternal loving perspective of understanding defines all things in one way, even from every vantage point? Does this raise a desire for GOD'S strong understanding in your soul? If so, this spiritual interest and pleasure are also part of the eternal perspective. We cannot know the first weak instance in which our eternal perspective begins, but we can sense it is now there with its growing indicators. I find my growing joy in the eternal perspective that always looks up; yet I get a very different feeling when looking at *this* contrary world. I see GOD'S eternal love in this world is used, abused, and denied everywhere in everything, by most everyone. To see this in most souls is heart-breaking; yet I am assured, firm, and anchored, to where *I* belong and desire to be. We see that as much as GOD is love, GOD is also the truth. The truths of life in each individual on earth, are all false, divisive, and will always fail. Sooner or later, they require being replaced by another false truth. No single godless truth of mankind's science, politics, philosophy, or religion, has endured the gauntlet of mankind's short history; GOD'S eternal truth has had no such problem. GOD'S eternal truth, started all things, but most all things have now turned away from that truth. Yet GOD'S truth endures, it has not changed, it cannot change, it is pure. The eternal truth defines all things minus paradoxes or required adaptations. Truth exists without ambiguities, vagaries, unfounded suppositions, paradoxes, impossible physical miracles, or strange meanderings. The eternal truth rests at and from before every beginning, reaching down from the highest Point, Pinnacle, and Acro, from GOD alone, who is its author. The first truth, must and will be the last truth, and come from no place other than GOD; for there is no other place from where it might come. Will GOD share their truth with us? Yes. This is the perfect perspective of the eternal truth. But those who are eternally lost, all they shall ever see, hear, or desire, shall not be real. Their individual truth will stay within their soul through eternity, to never be believed by another. Yet the saved shall live, walk, see, hear, and share together in FATHER'S created Heavenly reality forever. Do you love GOD? Then you should want more of GOD. Do you love the ways of GOD? Then you should want more of them. Do you love the wisdom and truth of GOD? Then you should want more of these also. This is in part with the eternal perspective. We must learn to sense all opposing forces projected at us from those who do not hold the eternal perspective. Even if their words seem to be meant as loving in their confusion, we sense subtle contrary forces within every worldly word. This is sorrowful for us, but more so if not sensed. Without doubt, to overcome this world requires our spiritual sensing and resisting of all obvious *and* subtle dark attempts at confusing our ways away from GOD'S ways. We must always put who we desire to be, up against who we in ways and at times might still be. In honestly evaluating who we are in each moment, we see agreements or declinations, and our remaining weaknesses. Our spiritual war comprises many battle fronts. We must confront and repel relentless attacks while out in this world. Most souls think they live towards a good temporary purpose, but they live against GOD'S eternal purpose. Our over-all sense of this spiritual war and its many battles, and our winning or losing, must always be known. We should always feel and know if we are winning or losing the over-all war, and our winning or losing in each battle. If we do not sense this, we are not fully spiritually engaged. This spiritual war in us all, has common aspects we can help each other with; yet we will have our unique aspects we must face with GOD alone. But in clinging tightly and only to GOD in true eternal love, there is no way we can lose. Our victory is set in FATHER'S created design to find and secure all such love. It is set within reality as FATHER'S immutable promise. Our way back was purchased by the great*est* loving suffering promise reality has ever known in JESUS, and by the comfort and teaching from HOLYSPIRIT who suffers to indwell us. And also, all *true* Brothers and Sisters in CHRIST, who lend strength. Once we are truly adjoined in all these facets, our desire for GOD pulls us on to victory. If we are *truly* and *fully* on the loving side of GOD, how can we fail? If all we desire is to climb out of this world, how can we not overcome all its illusions, seeing all GOD'S forces are aligned to secure us in love. But ever-so strangely, early on in this process towards eternal victory, some begin to sense this is not what they desire, and turn and walk away. And GOD cannot, and will not stop them. Again, we see "The Judas equation". We see that in some, their core desires remain as a hidden tenuous subtle mystery, even to themselves. They still hold a powerful worldly desire they did not overcome; and at their last, the stronger desire wins. Most souls lean more firmly one way or the other on different desires; but some can be divided nearly down the center of certain core desires. If you struggle for far too long with a certain desire, this indicates you are strongly divided in that desire. Do not think these cannot be overcome, because they can. But we must recognize these, and increase our effort in any area of danger. The loss of every war, has followed losing *one* big very strategic battle. We are the front-line commander in our own spiritual battle; and in loving GOD, we are invincible. If we in practice, assure our allegiance to FATHER, he assures us victory in him. FATHER, our words and our every form of expression fail to declare your full wonder and love and truth. HOLYSPIRIT, our comforting teacher in this world, you are our hope and brightness as you come to us in all measures from above. Holy SON of GOD, my KING, my bright eye of faithful responsible love. JESUS, who put off your physical *immortality* of GOD, taking on the physical *mortality* of mankind, putting your *immortal* soul at risk for us. You came, walked, and suffered, bound to the impoverished flesh of mankind. You came down so low to take and feel our every *just* horrific punishment of life itself, staying faithfully and responsibly bound in FATHER'S love. FATHER, you are before all things, being then, now, and eternal, the force of understanding, and great creator. Your highest thoughts are forever set beyond our knowing; the reach and measure of your love and wisdom astound us. Show us FATHER who we are to you, as who we are to ourself matters not. FATHER what do you see if you gaze at me? Do my understanding, love, and trust in you grow and affirm? Am I considered in your eternal realm of love? Am I a loving hopeful possibility? Or a rising loving probability? Will I ever be sealed as a loving certainty? Of all that has ever been known, and will ever be known, this is *my* great eternal question. A question only you oh GOD above, can and will in all your Righteous and Holy ability answer. I listen and watch, and as all my eternal questions arise, you have granted me your undeniable immutable answers. But for my answer to this last question, I must wait. This answer comes only when you see my oneness of understanding and belief. In Scripture, you tell us to *run* the race, so, I know there is a race to run. I have much to find, and weakness to overcome. I despise my weakness, as I fight to strengthen my love and perspective. Amen. Though we in great wonder, discuss the great truth and understanding of GOD without end, it is all a simple elegant thing to understand. It is bound in the perfect logic of our eternal loving responsibility and faithfulness. The only thing that complicates and blurs our understanding of reality, is our surrender to the temporary desires for the imperfect impure ways of life in this fallen world. That conscious fog of desire, and opaque-mindedness, that come from imperfect desires, is often impenetrable from the outside. We always try from outside to reach others, and can at times plant seeds of truth that may grow. But Scripture says to grow, a seed must first fall on fertile ground. Certain qualities and measures of Godly love, hope, and fear, must be present in a soul, for GOD'S love and truth to take root. But most lost souls walk around in the dark hell of false truths and beliefs, having set themselves beyond GOD'S loving truth. We must see, understand, and accept, this fact. It's the eternal aspects alone, that make GOD'S love so valuable, and meaningful. FATHER is a great vast firm and ethereal image, ruling over his great created theatre. And as FATHER'S truth shines from reality's foundation to its foreground, no matter how many times we say it is right here and there and everywhere; from their temporary perspective, most can't see or hear his truth. These can't even see how their temporary false beliefs are riddled with contradictions. There are many great common contradictions, and also the individual lesser beliefs the lost stumble over. HOLYSPIRIT can see and use these weaknesses in rare moments if *we* are there and led to say the right thing, at just the right moment; while hundreds of other souls could point out truths at so many various times, and it is brushed aside. Let us use a science fiction analogy and see lost souls as lost spaceships under what they sense is an attack, and they have their forceful protective shields up. So, they feel nothing of consequence from thousands of loving words projected at them. But with HOLYSPIRIT as our Captain, scanning when and where their shields are weak or down, and aiming a perfect loving understanding truth at just the right point of weakness, at just the right time, we may break through. Such rare opportunities in this ever-devolving world can fall upon us by chance; so, timing and decerning and wording matter so much. If in true loving desire, we say the right thing at the wrong time, we may only make things worse; we have all experienced this. It is no great mystery that there is the perfect thing to say, *at* the perfect time. I have been around many souls and wanted to say things, but did not feel led to do so. And in my zealous moments, I sometimes say something anyway, and it was a mistake. I have been asked by souls to talk to someone, but did not feel led; and the person asking can't understand this. For me, being led by GOD, is a strong feeling of unusual assuring pressure I have come to trust. And when I yield, there is peace, and the words do not sound like me. But when I speak from myself, I stumble and struggle to find the words. I am no orator or writer in any aspect; but if a soul is led, strong words come from elsewhere. In this, we too are listeners in most ways. I feel *I* am being moved, more than I feel *I* am writing or speaking. I know this beyond all doubt, as these moments often have elements beyond my ability or knowing. Words like ability, ken, or knowing, imply our knowledge came from being previously learned in this world. Yet I have come to know many things not learned from this world, that came in differing strengths at different times and in different ways. If we speak in faith, and hear ourself say something, but then think, wait, I do not know that, or how did I know that; yet later it turns out to be true. This is an enigmatic experiential spiritual-qualia GOD'S Children sense in their degrees. Why do we sense this in differing degrees, because we are all at our own distance from GOD'S perfection and strength. When I know, walk in, and share only GOD'S ways, beliefs, and understanding, I walk in perfection; and I am perfect in the eyes of GOD. If I stumble a step off that path, I have stepped into my imperfection. And if I sin, in my greater growing shame each time, I must again bow, and I must ask my KING to once-again bear the shame and stripes and pain of my unfaithful irresponsibility on the cross. This is a harder and harder shameful thing for me to do each time; as I cannot understand how I could betray the one I am supposed to love above all that exists. And I must ask, how can you my Holy KING, once again forgive my unfaithful irresponsibility to all that is good and eternal. Though I stumble far less often, I cannot forgive myself. Yet to date, if I stumble, GOD has returned to me; but it does take longer sometimes. In my most recent stumble, it was 4-5 weeks of shame, fasting, and prayer, before I felt HOLYSPIRIT'S return. By this, I know I have hurt and disappointed GOD'S love greatly; and I know I can't continue to stumble forever and eternally be forgiven. Will my love for GOD fall short at the last? How can I feel close and yet never close enough? Will GOD not return to me, if I betray his eternal love one more time? There came a point of no more forgiveness for Judas; and it coincided with the point of where his shame and confusion were so great, he could not even ask. All sin is our betrayal of eternal love and the ways of life created by GOD. GOD will use everything *available* in each moment to reach any in whom he senses a hope they might hear. A weapon that misfires every tenth time, is better than one that misfires every other time. By this analogy, if we are *not* led to speak the right words at the right time that HOLYSPIRIT sees an opening, to spiritually help someone, this *is* a misfire. In Scripture, GOD divides all the works or things we do into good or bad works. Scriptures declare that a good work is that which sustains, or in a way and measure, increases the Spiritual strength of GOD'S Kingdom. If one does many things, being led of themselves; things they hope will increase GOD'S Kingdom, but none do. These were then individual fruitless works, *outside* of being led in the will, truth, and timing of GOD, that more often strengthen the flesh, and move souls away from the *eternal* good. There is a profound difference between what is seen by mankind as being a temporary physical good, and what GOD sees as an eternal spiritual good. We see from the eternal perspective, that being led by HOLYSPIRIT alone, and not ourselves, or any worldly consensus, is key. To treat any *true* physical starvation, pain, or suffering, in this world is a good thing. But this alone, so rarely moves any soul from a temporary worldly perspective towards the eternal spiritual perspective, and often reinforces the temporary perspective. If you give and give food to the less fortunate. What does that mean? Are you doing a good thing? Are we called to feed the lost, or lead souls to CHRIST? I can feed a poor overweight reprobate forever, and not accomplish one Godly thing. If I say by doing this, I do GOD'S work, I misrepresent, overstate, and I am being legalistic? Yes, we are called to feed the hungry, but these being less fortunate, has nothing to do with it. Can any go to Heaven by doing physical works alone? Feeding the less fortunate is like Eve eating the fruit, it had nothing to do with fruit, it is about a soul's spiritual outcome. Do the first work first, and the second work second. In these last days of growing confusions of evil desires, it will more and more require the words and timing of HOLYSPIRIT to plant a deep seed of hope. Let us add GOD'S truth to every charitable work, and not give any physical abundance above the actual physical need to live. If one meets the temporary physical desires and wants and lusts one holds for storing-up an abundance in this world, this is not a Holy Work. The so-called poor of the advanced nations in this world grow obese, as they lose all spiritual awareness, being addicted to more food than they physically need. It has been said, that if we accept a sin in life, the next generation will revel in that sin; and the next generation will no longer see it as sin. The words want, covetous, and lust, have now become one in this world with the word need. If one wants something physical that another has, most don't say they want it, they believe they need it. Do we not see that the subtle creeping forces of evil physical sensations are in control? The forces of nurturing our physical desires, are detrimental to our spiritual desires. The ways of a physical, or spiritual life, are diametrically opposed paths, perspectives, and outcomes; they do not mix or blend. And both cannot be held in any abundant way at the same time. At their last breath of this worlds air, most souls end up being physically defined. No one can find oneness with GOD, yet be an individual. The desirous choices we all must make between being in GOD'S family, or of ourself and this world, are always right in front of us. In this world, these choices never go away. We cannot refuse to make or choice, or deny having made our choice; we make our choice all day long. Throughout Scripture, GOD says if we do not lovingly choose them *and* their eternal ways, we are choosing our self *and* our own ways. Is this not plain and simple to see in Eve and then Adam; there is no third choice. There are GODS true eternal ways, or each soul's desired illusion of temporary individual delusion. A truc choicc is sct bcforc, among, and within all souls under the one Creative GOD above, set to choose one of two eternal outcomes. And it is always and only between these two choices. To be allowed to choose between three or more eternal outcomes, is unreasoned and without purpose or perfection. FATHER will fulfill all his family's eternal needs. Individuals making many temporary choices, bring their own divisions of confused hell and destruction upon themselves. Satan, the father of all lies, is the father of all individuals, the father of all hell and death, and all separations from GOD. Satan was the first source and "Sower" of every lying confused illusion and delusion. Satan desires to change all he sees to suit him, yet he only destroys all he changes. In pure denial and delusion, Satan attempts to steal something to which he has no right or claim, or power to sustain. Satan can take nothing from GOD; but he and all his followers can lure souls to reject GOD and surrender to worldy individual desires and curiosity. All have heard that Satan is a liar, but who talks about his thievery. All that Satan seeks to steal and change, are the gifted souls, and the created firm perceptions of GOD, being the moments of GOD'S firm external shared dream. This once perfect reality FATHER created, is diminished by every evil desire; but the souls Satan lures into delusion will never be his eternal subjects. Satan, like all who rebel, at the last gains nothing, and loses everything. Satan will never rule over a kingdom of hell and death. Scripture says after all Judgment, all lost souls will narrowly look upon Satan in his weakness, being exactly like they are, after GOD casts him down into his final powerless form. Seen by all, Satan, sitting in his darkness, against the sides of the pit. All will stare upon him in wonder, saying is this he who destroyed it all, who is now brought down to be powerless and weak beyond description. Unable to be above anyone, imprisoned forever with those who will never have a single thing they desire. Such never-ending ever-growing conscious anguish and regret, for following evil desires that cannot lead them one step from their suffering. Can you imagine feeling nothing but confused growing conscious anguish and regret forever. Not one moment of joy, peace, love, comfort, or pleasure. And able to see that every choice they made was wrong over and over for eternity. No good things will ever exist there, for all good things have always, and will always flow from GOD'S love; and all good things are sustained only by GOD. This is the eternal loving work of FATHER for his Children. Please see this immutable faithful promise from FATHER, that he will eternally work to keep us safe and happy; *if* we eternally will not forcefully alter and diminish what he creates for us. Please do not think that anything as wondrous as eternal love and peace among many souls can come, or be sustained by anything other than the loving perfect ways of GOD and us together. The great complexity of such a thing, can never come, or be sustained by any below GOD. Thankfully, FATHER is a vast loving Creative Soul. Do you feel any aversion to, or unfairness, in GOD'S designed ways of life. Do you wish that everyone could design their own personal heaven, so their contrary ways are seen as good and eternal before FATHER? This cannot and will not never happen. A temporary perspective does not disappear in an instant when turning to GOD, re-building our believed understanding is a true process. If any ventures many steps away from GOD, they must turn and undo those same many corrective steps back. Or, do any think FATHER should do that too; that FATHER should bear even our works of giving and showing our love for him. Should we receive all loving works and gifts from FATHER, and there be no true or real loving gifts or acts performed by us for him. The greatest true aspect of *eternal* love is found in its shared eternal flowing circular measures. How many times have the loves in this world failed us, how many times have we seen love diminish, die, and often turn to hate. I tell you a great truth, always love and cherish GOD more than anyone or anything, and you will always be spiritually loved nourished and held by GOD; and you will hold an eternal love for all others of like mind. If we hold our greatest love for those souls of our biological family, that equation brings its own divisions of hell, and death. We must love, cherish, and nourish GOD'S Children, as much as we love the life within us. There are only two eternal families: FATHER'S and no one's. We hold great hope for love in all souls when we first meet, but this hope rarely gets to grow. This is both, our aspirational *and* sorrowful understanding. Many souls take a few steps up, but then look back down at the souls they love who do not step up with them, and feel a forceful pulling back by the love they have for the souls and things in this world. They find a stronger need for feeling and focusing their love upon their biological family, above the spiritual family of GOD. JESUS said in Matt. 10:37, "*HE THAT LOVETH FATHER OR MOTHER MORE THAN ME; AND HE THAT LOVETH SON OR DAUGHTER MORE THAN ME IS NOT WORTHY OF ME*" Most souls desire their mortal family more; our love for our mortal family is not evil idolatry, until it exceeds our love for GOD. And our love for our worldly family secures nothing eternal for them or us. If one gives up trying to find more and more of GOD, the only thing left to do, is to blind oneself with more and more coping mechanism of false beliefs in compromised ways, things, and family. Souls can easily fill their life with the illusions and delusions of worldly family, friends, belief, and religion to help face each day and each other, in the denial into which they sink; but why? They refuse to see or focus on the soon approaching moment of truth; but why? And most achieve this self-delusion to incredible levels, and many revel in it; but why? I assure you; all prideful arrogance is temporary. Many deny it all, saying they have physical scientific beliefs, as that becomes their desired belief. When looking upon the vast fields of losing lost souls, who surrender to be among the doomed, it astonishes and grieves us. But what astonishes and grieves us more, is that in their few decreasing moments of spiritual reflection, most blame GOD for theirs and this world's failures. If at the end of my choosing I should fail, I pray I forever know it was me who failed in my insufficient desire for the loving provision, ways, and understanding, of GOD; as that would be my small eternal measure of sanity. GOD has never, and will never fail to be righteous, holy, and eternally loving. If I fail, my regret should be the greatest regret of all souls, for coming close to see and know GOD and yet fail. A feeling Judas, the betrayer of CHRIST knows well. Dear GOD, to my very last choice I cry against the dark, and grasp for your beauty, reason, and purposed ways of love. Consider how the endless antichrist forces in this world work so hard to move all souls and their evil ways. Antichrist forces rule all false liberal Godless immoral politics, all false liberal confused media and education, all false liberal confused science, and all false liberal religion. And now they are moving into policing and military command positions. These grow and congeal each day in a new world order of antichrist design. Many see politicians as a huge political force, but they are but tiny rudders of huge ships. Those who acquire a louder voice, just steer and focus gathering forces. These evil forces gather in impassioned desire. When these blended *evil* forces swell, they breed a mass psychosis of which some are worse than others. When conscious forces gather together, a political artist funnels their passion into a compromised point for the many to see a greater good on their side; and a greater evil on the other side. All who do not follow GOD, can only follow what they perceive to be the lesser of two evils. Can we truly follow GOD, and vote for the lesser of two evils? Yes, as long as we know we are only slowing evil, and not correcting it. Dangerous liberal forces, place an *evil* political artist at the helm. Remember that Hitler's confidants in the media, lied to demonize Jews for several years. And as the numbers who believed those lies grew, the Nazis created a mass psychosis among German citizens who cheered as Nazis rounded up millions of Jews and gassed and burned them. Do you not see this world's demo(n)cratic political education system, science, media, and immoral religion, demonizing those who resist their forceful work to liberalize and redefine reality. The political artist has two hands: one grabs hold of the masses in the light of day, while the other hand steers that power in the secret of darkness; these are the liberal antichrist ways of this world. Throughout the ages there has been a global antichrist agenda at work. Once working more in the darkness, deep behind the scenes; but now they openly rail and create many antichrist individuals to gain global control. While in this world, these are Satan's Children who see the world through the vision of a godless ruling class. These know how to use every divisive word, weapon, and dollar, to their full extent. And liberals through the deep-state of bureaucracy, control most public power and dollars. Satan through these deep state means, works to control the world in all its disillusions, by separating everyone from GOD and each other. It is the bond between GOD and Godly families, Godly communities, Godly States, and Godly Countries, that stops Satan from uniting a worldly consensus under one confused godless rule. Satan's children work to weakening and break every true Godly bond, as they know they must destroy all eternal hope in life and love.

Liberal hypnotists say they will do good things, while doing only pure evil. By controlling the lying media, they project and propagate their confusions of hell upon all who do not know GOD. And souls caught up in this mass psychosis, believe most anything these antichrist propogandist say. Liberal leaders try to stay a step ahead, to keep their base riled up in physical resistance, and keep them from seeing the truth. Make no mistake, this is the *Spirit* of antichrist. "Spirit" meaning; attributes, beliefs, desires, ways, and efforts. And "Antichrist" meaning a movement aligned in one purpose, being against the perfect ways and commands of GOD, for control of the earth and power over all who walk upon it. Such lying thievery; and at the close of this third age, this disillusioned world will en-masse, unite behind a self-proclaimed insane false liberal god, who accuses the true GOD of being false. And the last insane goal of the masses, will be to track down and publicly convert or kill all true children of GOD. This is Satan's final goal in Scripture's last days Prophecies. This will soon come against all *we* love most dearly. How long until this occurs, I would say at the earliest, this actual hunting *could* begin in twenty five or so years from this writing, at the outside, not a whole lot longer. This will be declared to be a hunt for the enemies of the peaceful world order under a false god. We will be hunted, framed, labeled as bigots, racist, fascist, and terrorist. They will commit every form of evil, and frame us as evil without hesitation. As prophecy unfolds step by step, we will gain a more accurate timing. So, the next *Prophesied* event, will be a war bringing the defeat of Iran. I know that most who falsely stand in pulpits, caution all to never offer timelines in GOD'S Prophecy. But Scriptural Prophecy says to observe, warn, and yes, offer appropriate timelines. If a weak preacher says we are in the last days; is this too, just a broader *safer* reference to a time-line of occurrence and happening, a weak ungodly preachers should not make. The reason GOD gives us Prophecy, is to let us know his approaching intercessory events that will bring a cleansing of evil. I have kept the bulk of the physical scientific understanding of GOD'S created reality, out of this particular book for a reason; but I must oft refer to GOD'S omni-science, as the eternal perspective uncovers another ongoing lie of Satan. As I said in another book, in all of this world's scientific theories, they keep running into unexplainable adversarial problems that require imaginative leaps of faith on their part. And we remind you that all their theories at their start, require a physical miracle. They hate this word miracle, as it is strongly associated with

GOD, so, they create new terms that have the same meaning, yet they imply they don't; like the term "strong emergence". When confronted about this term, most scientist try hard to dance and talk around its core meaning. Because it means the sudden appearance of something beyond the powers in the chain of events of physical cause and effect. Something new, *not* associated to any prior physical chanceful events. So, *we* ask, what conscious effects are they trying not to explain. We know the word "miracle" means an occurrence, contrary too, and outside of mankind's modeled understanding of physics, but so does strong emergence. Yet once we intimately know GOD, all things are explained as desired conscious creations, and the term "Miracle" loses all higher meaning. Under GOD, there are no unexplainable occurrences. There are only his and our consciously created complex perceptions formed of particle wave properties; and every perception has been defined by the forceful desire that created it. So, if science sees a miracle, meaning their model of reality does not allow it, and can't explain it, then everything that occurs following that miracle, is also unexplained. The sciences can't slip in deceptive miraculous occurrences here and there that happen for reasons beyond all cause and effect, and declare all that follows to be a natural occurrence. It is beyond question, that the believing of all things by every soul, is never found on our surface; but deep down within our sphere center of being. But mankind's perceptive ability, perceives and shares only the surface of everything. Mankind has peeled back a few surface layers here and there, but they have never seen or known the true core aspects of reality; therefore, they do not truly know anything. They cannot write the first letter of the first word of the first sentence, that in any physical way, can plausibly explain, or begin to prove the first cause of the universe, or life, or conscious awareness, or free-will agency, or building and destroying complexity, or time and eternity, or force itself, or anything at all. Pure physical science, can never know the countless smaller things in between the far larger things they do sense. Again, there are only two known aspects to all reality: one is every soul's forceful patterned desires that move, build, or destroy. And two, being the purely physical aspects, which are then moved, built-up, or destroyed. If a single aspect of any occurrence, cannot be defined as that which was created or destroyed, it is then an aspect of that which creates or destroys. And the latter being the intentional generated and projected forces in our desires to exist. Now again, the tiny measures of *our* souls have *eternally existed* as a former part of FATHER'S soul; it was not new or created at *our* first awareness; it was partitioned as clean and clear, without prior desire or understanding, and gifted to become us. And *our* first awareness and *eternal* ownership of our gifted soul, did begin. So, *our* awareness and chosen spirit, *are* a new reality. And all *our* desires, are forceful influences that create certain aspects, and affect other physical aspects that fall within our reach of force. Now the closest thing to a miracle or strong emergence I have ever seen, in knowing GOD is there; are destruction and evil. Because destruction and evil can only appear *after* GOD'S perfect creation. Destruction can never occur before creation, there being nothing to destroy. In GOD'S conscious primacy, all physical imperfection, meaning all things contrary and destructive, can only enter after and come against something perfect. Imperfection can never exist prior to perfection and have meaning. Now before FATHER'S external creations, and gifting many eternal souls, he was utterly alone. So, there was nothing that occurred beyond the spherical internal created substantive particle-wave-based, desired dreams of FATHER. And all that was perceived by his desired forceful cause and effect within, and eventually external to FATHER, were all perfectly perceived, or created. After FATHER'S perfect external creation, and dividing out many free-willed souls, Satan in his free-will desire, was the first to choose to become imperfect and individual; being forcefully contrary to the perfection of GOD. This was the *closest* thing to an emergent that has ever occurred from the perspective of GOD, as he did not choose or cause it. Here something first occurred completely beyond the omni-scientific cause and effect of GOD. But this emergence, still came by another free-will force of conscious desire. Making it an explainable conscious force caused physical effect; and no true physical emergent. Yet in a way, it is a strong conscious spiritual emergent in the mind of FATHER'S Family, as it makes no eternal loving sense. Now further, as to weak emergence, we all easily see that nothing weakly emerges, that term meaning just simple cause and effect. So again, we see a contrivance of a new term, meaning nothing more than physical cause and effect. We see, they could not deceptively claim there is a strong emergent, without claiming the useless definition of weak emergence. In all of GOD'S reality, there is no such thing as a physical miracle, or physical strong emergence. Even if every individual desires to take such an unbelievable and indefensible position. Such false obscure beliefs are born of an unfounded godless faith. The term weak immergence, has no unknown stage of occurrence, it does not sit alone between two different aspects; it's a term void of meaning, yet it has its evil purpose. *Many* words in secular scientific nomenclature, are no more than misdirection to turn souls back into confusion, lest they get near enough to recognize a conscious spiritual truth. Think of these terms as designed false information exits or loops, cleverly used to keep souls from climbing out of a false physical scientific faith. These false phrases and terms seem obvious to us, but to those wanting blind spiritual darkness, they walk among these lies all the day long, resting in their faith that they are true. So, which is worse, physical blindness, or spiritual blindness. I can see where physical blindness might even be a blessing for some; but spiritual blindness is the greatest eternal curse a soul can inflict upon their self. Please invest your time to study that new term strong emergence, and ask yourself if you see any difference between their definitions of strong emergence, and miracle. Though they try to imply there is one, there is not. If you read any scientific proffer that uses the term "strong emergent" just replace it with miracle or miraculous, and see what they truly mean. And again, when considering the cause in cause and effect, we must dig deep until we find the very first cause; being a *newly* generated projected creative force. Once a force of desire is projected, we see that force can be stored, carried, and transferred. But at its beginning, before any stored, carried, or transferred force exists, where did that force come from. Force has been *newly* generated and projected in patterned desired intent for all eternity. Again, eternity is spherical, ordered, immutable, and known; not linear, wild, unknown, and dragging reality towards destruction as time does. So, a first generator of new force, must have existed prior to all acting forces of intent. And we find no source, other than our conscious desirous FATHER, desiring and intending life to exist and love. We all generate new, never before existing force. But materialist say, all force is accounted for as being once stored and released in a Big Bang; and again, without ever declaring what first generated that force. Yet the forces we do see, refuse to be reverse engineered back into a big bang. New force, appears in every created conception and perception. And conception means there is a conscious beginning to all reality, and reality is a consciously forced continuum. The properly understood sequence is, each soul's new willful desires generate and project mirrored patterns of forceful intent that create. And with many free-will souls, many projected *contrary* forces, are followed by currents of destructive physical effect, and so on. Now a question I hear, is how can we walk away from certain souls we love in this world. But we say, we walk away from no one. We walk with and towards GOD as those souls are all in our heart. And if distance grows between us, why is it we, and not they that walk away from us. Liberals always sees themselves as a victim, and all opposition coming at them, and never from them. This is insane narcissism and delusion at a minimum. We walk the true way, yet they blame us. The first instinct in those who abandon GOD, is to accuse others of abandoning them; but why? Our only other choice is to leave GOD and follow them into hell and death. I have left no one behind, just as GOD does not wantonly leave any behind. It is always the other way-round. But this eternal truth is only seen through the eternal perspective, as all temporary beliefs are seen through temporary perspectives. If I, for my worldly family, would turn back into temporary physical confusion, to be closer to them, I would abandon my responsibility to reach and intercede for them. I love my neighbor as much as I love myself; but if I loved them more than I love GOD, I would follow them, and not GOD. But I love GOD so much more. I treasure FATHER'S gifts of their, and my eternal soul equally. I long for all souls to treasure, cherish, and nourish, their eternal gift. If *any* soul throws the great promise of their eternal gift away; this can raise no reason, or purpose, or loving desire, or eternal wisdom in me, to do the same. And I know with all absolute resolve and certitude, that what I do is right. And this I must do, as this is who *I* am. Do you follow your loved ones into their every dark destructive behavior. How can any but a confused dark soul, do such an evil thing. I do not say that the forces that pull most down, do not seek advantage against me also; but only, that in my awareness, I see and know their end result. So, I desire to resist them more than most. And surely no one can defeat an enemy they do not want to see or understand as being their enemy. Ever so sadly, many do not see that they actually love and aid their enemy. They cling to, and run into all conscious individual divisions of destruction, hell, and death. They think that all contrary things and beliefs can be had and exist forever as good, not seeing they only destroy one another. We are created to live in GOD'S righteousness and holiness in each moment. Let the moment change, but not yourself. By evil core desires and need to be an individual, and falling too far to return; these unknowingly dig their spiritual graves and enshrine their soul in eternal hell and the Second Death. The Scriptural first, and second death, literally mean the first temporary, and then the second permanent separation from

GOD'S provisions of love and sustenance; but why choose death? The closer one gets to GOD, the less one understands rebellion in others. And the further a soul gets from GOD in rebellion, the less they understand our loving obedient desire *for* GOD. Do you see this final conscious spiritual divide being formed by all free-will. An insurmountable chasm is opening between these two contrary perspectives; and souls form this void of their own volition. Just as great brightness and utter darkness cannot occupy the same space, by free-will, this just must be. And all that will be required at the last, is for GOD to seal, and perceive two eternal places for these two families of souls, that can never again understand or know each other. Two places of contrary perspectives that can only forever gaze upon each other in wonder, as they spiritually grow further apart. Those in Heaven gaze over, knowing and understanding all truth and elation, feeling the *past* painful residue of lost loves, forever asking why those souls resisted the perfect eternal estate of love among many that we will forever experience. And the lost, look over at us, having no love or provision, forever separated from GOD/ Trapped in a conscious state of their own confused design of desire and false understanding, completely individualized in a conscious state of hell. Experiencing all *conscious* loss in death's separation, and hell's separation, they are cast into the inescapable physical lake of fission and fire. It being an eternal prison of accountability for all evil, and a loving protection for the good. Rev. 20:14-15 "*AND DEATH AND HELL WERE CAST INTO THE LAKE OF FIRE. THIS IS THE SECOND DEATH. AND WHOSOEVER WAS NOT FOUND WRITTEN IN THE BOOK OF LIFE WAS CAST INTO THE LAKE OF FIRE.*" When speaking of this prison of fire. I remind all, these souls are first given eternal bodies, which cannot be killed or consumed by fire. So, these bodies have far less physical feeling than the ones we now occupy. MARK 12:25, "*FOR WHEN THEY SHALL RISE FROM THE DEAD, THEY NEITHER MARRY, NOR ARE GIVEN IN MARRIAGE; BUT ARE AS THE ANGELS WHICH ARE IN HEAVEN.*" Isaiah 43:2, "*WHEN THOU PASSEST THROUGH THE WATERS, I WILL BE WITH THEE; AND THROUGH THE RIVERS, THEY SHALL NOT OVERFLOW THEE: WHEN THOU WALKEST THROUGH THE FIRE, THOU SHALT NOT BE BURNED; NEITHER SHALL THE FLAME KINDLE UPON THEE.*" In this prison there will be no pain from GOD. The level of physical feelings there, may feel on the very warm, dry, thirsty, and uncomfortable side. In this prison, for a soul to move one arm against its immense gravity, will require great effort. Any hope for a tiny physical pleasure or comfort can never come. All conscious sufferings of anguish and regret will be justly experienced to differing extents, all according to the evil confusion within each soul. Will the reward Heaven be measured out to souls differently, according to their lesser to greater levels of created good? Yes. But they will be experienced on different levels; so, this matters not, being an experiential shared quale or qualia of all being in the perfect spiritual creation of love and peace and life eternal. Heaven can only bring as much joy to each soul as we can hold and experience; so, we can never know or care if another experiences more joy. Now the individual endless thoughts of division in *this* world, appear as tiny local distortions of forced disorder that contribute to this growing divide in all conscious reality itself. We see growing cracks of separation between individuals, families, and countries. And where ever fractures appear, they usually grow; some slowly, some fast; a few may appear to stop and recede. But unless these divided souls turn and go to GOD, these cracks between them cannot heal. Most fractures start and grow below the surface unnoticed, and people are surprised when a sudden brake between themselves and others occurs. Souls say, I did not see that coming. A temporary perspective focuses on the surface appearance, as they more and more see only the near and the physical. They rarely look into the core spiritual aspect of life where cracks first appear. With cracks, a weakness appears; so, a cracked foundation is followed by cracked walls, ceilings, and eventual collapse. Begin to build your soul's understanding and loving desires upon GOD'S ways, truth, and belief, as your foundation, and stand together for eternity. All life's meaning, rests in the words perfect and eternal. Now at first, an eternal perspective can feel uneasy; at my eternal perspective's beginning, I felt a bit divided inside. But I've grown to know all truth and meaning are only found in the good desires of GOD'S eternal family. We exist in the eternal current moment, as we always will. I know we all will consciously exist without end, as each soul came from FATHER; a part of him who cannot cease to be. I know we can be joyous in every moment that is now, when all reality is provided by GOD. I know and trust in that immutable promise of GOD, that they know how, and are able, and will provide us an eternal peaceful way of life. We see it must be a perfect way of life, designed and provided by FATHER; and we must rest in our following GOD into that life. Do not be confused by, or afraid of this, but desire this! Do not be foolish and rebellious towards this, desire this! Be set in your greatest loving hope for it, and your greatest fear of rejecting it. In this way, our joy of life is eternal as we allow, learn, and adopt our Godly eternal perspective of life. Again, at first, one might feel reality is now strange. Certain movies come to mind about people waking up out of a false reality, and into a real yet undeniable reality that takes a moment to get used to. But does the true reality feel strange, because it is strange. Or because we were so confused and convinced by a strange belief as to what reality is. The true reality can appear strange at first, but only in contrast to a lie of reality. Once we accept the reality of GOD and experience its perspective for ourself, it appears and feels proper and perfect; and that former false reality appears strange. By desire, we grow to be comfortable, accustomed, secured, and at home in GOD'S true reality. If your desire for GOD'S ways continue and grow, and you live through the eternal perspective for just a year or two, you should feel right at home. I hope and pray you will never want to go back. I actually don't see how I or any could turn back once the vale is lifted. Can one just cover again one's eyes, forget, and be re-deceived again by the same former lie, being a most dishonest conscious contrivance of self-delusion one could ever imagine; yet we see it happen as desires for their former ways return in some to become powerful. Do you sense the infallibility of the eternal omnipotent perspective of GOD? Do you feel and sense the forcefulness of GOD'S omniscient truth upon creation? Do you see that GOD cannot fail at the last? Do you grasp that GOD can create all things but love? Do you hold a true desire to explore this? Do you truly desire to experience this? And if you say yes to all these, the next question is why do you say yes? GOD draws us by good loving forces like a magnet, for those who knowingly and even unknowingly want and seek love and life without end. It is a forceful bonding attraction that exists between all like-minded forceful souls. Our eternal perspective of life, must be sensed, nurtured, and cherished, as with all the things of eternal life and love. View reality in the way GOD sees it. In his eternal design, it was never meant for souls to build their own individual false understanding and belief of reality. GOD'S eternal perspective and one pure truth are what we are meant to learn and understanding from our creation. This is the bond of eternal oneness of belief and purpose that exists between FATHER, his SON, his HOLYSPIRIT, and all his Holy Angels and Saints. How can any say they truly believe in GOD, and not believe their one pure truth and belief are not a foundation and prerequisite for eternal life with them?

Each soul must want to perpetuate life together as one, forever. Such a desire can never be forced upon, or into one's free-will desires. Coercion cannot occur without spiritually destroying who one freely is and desires to be, resulting in confused souls, like the liberals create. Why would a loving GOD, or any loving soul, desire such a thing? What could a loving soul needing love and life eternal, derive from such forced actions? Only a selfish envious demented soul desires such control over others, as they see their forced control of others as love of self. Satan's angels, and minions of godless liberals in this world, in all their delusions, desire these false things. They seek to possess and coercively bend the will of all to theirs. These want in hateful envy of GOD, to steal all souls through emotional controls. This desire permeates the Demo(n)cratic Party. And as all birds of a feather do flock together, those blinded against the true light, use fear, hate, and unbridled physical pleasure to lure, confuse, and entrap souls. Why do liberals instigate violence against the Godly? Because they hate the Godly. Why do they seek to sexually confuse the young who do not yet know better; because they sexually desire them, it is called grooming. Their greatest pleasure in life is to defile the pure and innocent. Imagine a most powerful unseen spiritual hurricane of evil raging all around you. When GOD opens your eyes to all the evil in this world, this is what you will see. Below the bright and shiny physical façade, there is a raging whirling spiritual darkness of anger, hell, and destruction, tearing at every innocence, and at every frayed edge of reality, seeking to break down all defenses. And in every battle, if defenses fail and fall, entry and control by the enemy follows. The naïveté of a temporary perspective, allows the temporary appearances to win. Some think GOD is losing, so, he must be the weaker. But we have expressed in so many ways, that for finding true love, GOD must allow all choices to be made by all, even blasphemy, murder, and rape. But at the close of finding all his love, FATHER need only glance and desire to imprison evil's physical side, rendering them helpless. FATHER need not even whisper, and every evil that has ever existed will cease within his reality; all evil will become powerless and of no affect. Not by controlling all thoughts and desires, but by greatly limiting their control over a body and environment that are not theirs. So, the lost, forever being in GOD'S lake-of-fire, that contains all their projected forces, and their new eternal body which barely moves or feels, greatly diminishes their abilities. The embodiment of physical pleasure they treasure most, is then revealed to not be theirs in their prior assumed ways. The new body they indwell, will be a never-ending, hardly feeling, barely mobile, prison. Now do not think the young souls who were murdered, including the half billion souls murdered in the womb, will be precluded from making their own final choice. GOD'S justice will allow for their final choices in CHRIST'S Millennial Reign. The great purpose of the last thousand years, is to settle all remaining injustice, and finalize all accountability. All souls who have not qualitatively found and made their eternal choice, in their required firm chosen way, get their required moment to do so. But we don't know if any new souls will be introduced there. If a woman had an abortion in this life, that child's soul will re-enter this world in CHRIST'S Millennium, through another woman. *All* souls must reach their fullness of desire and choice prior to their judgment. As with all Godly things, CHRIST'S Millennial Reign has its perfect reasons and its eternal purpose. GOD is truly just in each and every perfect accounting. Not like the corrupt mess of improper justice and unaccountability found under mankind. By GOD, no one will be overly-rewarded or too harshly judged, or be under-rewarded or judged too lightly. GOD'S judgment will be made according to our own core desires and chosen ways that define our life within GOD'S reality. Free-will choice is such a deep consideration; there are many forceful deterministic aspects and affects to consider, being projected in our every choice. Most consider and see the exterior reality in a false four-dimensional way. We must see and interpret reality in the proper six-dimensional Godly way. We must know, each sin, sows its imperfection, disorder, division, and destruction, as it turns GOD'S love away, and hurts his creation. All sins are contrary to the created eternal ways of life. Most seem unable to live in the eternal ways established by GOD before the beginning of creation. The worldly religious don't see the difference between GOD'S way of eternal perfection, and their ways of imperfection. Most think GOD'S truth is like theirs, malleable, and not yet finalized. Most assume they will in many ways, choose how they live in Heaven, as they do here. Open your eyes to the most basic understandings of reality, being souls see and interpret all things through their great desire to do so. If anyone finds this to be a complicated matter, by their desires they have made it complicated within their own soul. And it is then they, if they want GOD'S love and help, who must uncomplicate this within their soul. Go through your entire understanding like a storage unit. Throw out every understanding and belief that has no eternal perfect purpose. Being most beliefs beyond basic math, spelling, and original grammar (no slang); then hear and learn perfection from CHRIST. You may then grow to attain the mind of CHRIST. 1st Corinthians 2:16, "*FOR WHO HATH KNOWN THE MIND OF THE LORD, THAT HE MAY INSTRUCT HIM? BUT WE HAVE THE MIND OF CHRIST.*" I have written before, how I hate the term intelligence, the world subtly uses to describe our driving core beliefs. Yet, I must use this term here, to best allow me to convey certain things to many souls. Now what does it mean to be a Godly moral intellectual? Well, seeing public intelligence and morality are so subjective, we must yield to the Biblical interpretation of HOLYSPIRIT, for FATHER'S one true moral intelligence. We then see what the world should call, not a worldly, but a Godly intellectual as our subject. Being a soul holding a deep spiritual/physical understanding in being a minute creative God. And who being such, strives to never exaggerate or understate a single point in any *Godly* conversation, as any exaggeration or understating is lying; not to include *admitted* estimating. But the fallen who constantly exaggerate and understate, do not see this as lying; they call it stretching the truth. And by most believers in this world, GOD'S truth of morality is stretched into individual subjective lies. And the religious body, agree with, and revel in this practice. Every true moral intellectual, sees stretching the truth is as much a lie as any other. If you know you have five dollars, and you tell me you have six, is this a lie or an exaggeration. Surely you must see two words without distinction. Exaggerating and understating, are evil contrivances of spirit, formed to allow subtle dishonesty in communication. Now many will say, whether you say you have five or six of anything in this fallen corrupted world, is insignificant. Yet in the perfect unfolding of Heaven, it would be catastrophic. If a Politician, Scientist, or worldly Preacher are prone to exaggerate, or understate certain points, how can we trust anything they say. Playing loose with facts has become habitual and expected. All souls now sense a distrust in this world that can never be untangled; so, they just want to move on as best they can. Yes, this is part of a temporary perspective. The pure truth of GOD can never ever be conveyed in any tiny exaggerated or understated way. By one exaggerated or understated word, meaning becomes impure, and a subtle lie. Most souls are prone to exaggerate and understate, making it impossible to communicate in this world in any pure way. We cannot know *exactly* what others think and mean if they speak in less than pure honest ways. Mankind's communication is dying, and individual desires, beliefs, and lies, are the cause. We should never exaggerate or understate anything. Lies only corrupt distort and diminish communication, till the truth is no longer valued by, needed, or heard by this world. On high, right beside the bonding force of eternal love, the bonding force of the pure eternal truth is still there; while imperfect lies and hate go hand in hand in this fallen world. Only GOD can assure our salvation. Do I always communicate minus any slight exaggeration or understating? No. But I only get pulled towards these, when I try for too long to communicate with souls of a dishonest temporary perspective. So, I see why Scripture says not to spend much time on such souls, and to flee. Trying to communicate with reprobates, pull us towards them; we will only go closer and closer to them in trying to reach them. Scripture says we must accept that most souls do not in any way, want what we know and have. Never think that all souls can still be reached; for they can't. So, a greater aspect of reality to know, are the conscious force influences within and upon us all. Most souls are mesmerized and set in their belief by the chain of events that occur between the physical things in reality interacting with each other. I ask all, to in awe, consider the strong detailed clarity of your own created perceptions. Think about how you create your multi-sensory, 3D, rich, deep, particle-wave based perceptions; and please don't think you do this with wet brain meat. Ask any neuroscientist to show you where we all create perceptions, inside wet brain meat? Ask them if our ethereal perceptions are made out of quantum waves of energy and particles like everything else? Ask them how wet magic brain meat can create complex rich multi-sensory streaming structures, that in all our sensing, appear to exist and reach far outside the brain and body? And how are we aware of our perceptions through that wet magic brain meat? Ask what process occurs between the patterns of light stimulating the eyes retina, and the perceived brightness of *our* awareness of created visual perceptions? Ask, do we see reality first hand, or do we only see inexplicably bright echoes of all the external darkness, we have actually created? Ask them if physical light has actual bright properties, or creates our brightness? Ask, is the universe bright in its own existence, or is it only bright in our conscious awareness of it? Ask them if brightness is an aspect of conscious awareness alone? Ask, if the moving energy and matter in the sun and all the stars and planets and the entire universe are as black and dark as can be imagined? Ask them if outside of our created inner brightness, is light just utterly dark fast-moving wave patterns of dark energetic particles? Ask them for a step-by-step theorized process for how, why, and where, our created perceptions, followed by our bright awareness, occurs within magic brain meat? Ask these questions, and watch them freeze at hearing a long list of questions for which they have not one plausible answer. I tell you they cannot give you the first step, or the last, or a single step in between. They can and might say they believe in certain miraculous things; but that is all they can offer. Their beliefs are a faith, taught them by godless preachers in godless religious temples of lies; they have not the first proof of their beliefs. They have only a certain consensus standing together to imply, and expect us to infer, that they *know* many things we do not. And in that they can never truly know *these* things, from a slow small temporary perspective, their only real purpose in this world is to deceive children into believing there is no GOD. What a great price these must soon pay, and all who choose to believe them will also pay a never-ending price. Yes, the unrepentant teachers of such lies, will pay a manifold price. We know that at the gifting and creation of our bicameral sphere of existence, we are given many things of choice from GOD. But our key to *eternal life* is found in perfectly aligning our choices and beliefs to GOD'S choices and beliefs. Why do most neglect and lose, and so often abuse the lasting things of love? Now as to the dual foundation of reality, there must be a perfect ratio for our physical and spiritual sides that allows many to live together in love for eternity. For so many to live for eternity in love, it is most evident that we are called to be far more spiritual, and far less physical. So, there must be a minimal threshold to perpetuate this. We know it is not a low threshold of 49% physical to 51% spiritual, it must be far more spiritual than that. What is the perfect ratio? This is not fully revealed to us, but we sense it is around 6 sevenths spiritual, and 1 seventh physical. We should see all things through the prism, that if our Creator adheres to it, being eternal, we need to adhere to it. Now there are physical things we must through a body do, that GOD does not do. For example, GOD does not need to, yet the body we use, must process and circulate energy. So, we must see our perfect latitude of choice within all this. If two foods are presented before us, and one tastes offensive to us, we can surely choose the other. And if both are pleasant, but one more so; we aught choose the lesser. Now so no one runs to take our latitude of choice too far. We see GOD created two physical halves of mankind to choose among, to procreate; which GOD also does not do. And this perfect physical function, was created for procreation only, *not* recreation or pleasure. So, any congress between souls, other than men and women, married in the word of GOD for procreation, range from acts of godly spiritual abandon, to abominable vile sin. Yes, GOD says sin is sin, but all sin is not an abomination. GOD says different evils exhibit differing levels of confusion and destruction; and perverse sensual sin is among the most evil and destructive. If FATHER in Scripture calls something an abominable perversion of his created ways; it sits among the highest sins, beside blasphemy, murder, and rape. Now let's consider certain aspect of our day, and sleep dreams. Upon awaking, and remembering a sleep dream, I consider and focus upon the strengths of reality within that dream. As there is evidence in the levels and depths of sensed conscious emotions and rich details of color, sound, touch, smells, and tastes. These lend insight into the sources of information we use for our soul's sleep-dreams. This is important, because we indwell both our wakeful and sleep perceptions, which are *all* formed in a dualistic fashion. Our wakeful perceptions rely mostly on the abundant strong information the body relays from all contributing worldly sources. And just as we need to know if the *sources* of information, we sense in our wakeful perceptions are evil, we need to also know this for sleep perceptions. For all sensed information, it is important in dualism to know its source *and intent*. The information we use to create our dream perceptions, comes from somewhere. So, it is helpful to know every source of information, and its *intent*. The projected information from all sources is good or evil. All information I sense from all conscious/physical sources, influences *my* perceptions. And if I cannot know and judge a conscious source, I cannot know their intent towards my soul. Sleep dream information, can be just weak memory remnants from within the aether filling our spherical soul; or from a Godly quantum realm soul, trying to communicate meaning of good intent; or dark souls having an individual perspective, having evil intent. 1st John 4:1, "*BELOVED, BELIEVE NOT EVERY SPIRIT, BUT TRY THE SPIRITS WHETHER THEY ARE OF GOD: BECAUSE MANY FALSE PROPHETS ARE GONE OUT INTO THE WORLD.*" Also, 1st Corinthians 15:33, "*BE NOT DECEIVED: EVIL COMMUNICATIONS CORRUPT GOOD MANNERS.*" We are not wakefully aware, or privy to sources of information we sense in sleep dreams. In wakeful perceptions, we are far more aware of sources, and keenly sense and know the firm very rich information of the creation around us. We strongly sense patterned information from souls within our reach of sight, sound, and touch, all according to their forceful words and actions. All such information, brings our rich perceiving, indwelling, and our correct or incorrect interpretation of it all. But when our souls lay more detached from our bodies in a dark quiet room under warm coving, that information from worldly physical sources, tends to be a negligible contributor, as quantum realm sources of information can feel stronger to our soul. We do *not* discount vague faint quantum aether memory and the body's neural response to dream emotions as contributing to our sleeping sensed information. Yet if other souls in the quantum realm, where all souls dwell, come near to our soul as the body is turned off, and they overlay or touch our soul to share certain forceful information we may call a vision, having a good or evil intent. And according to that information's strength and detail, our soul may be charged to very clearly perceive another soul's projected information. Our own internal dreams tend to be weak and intermittent, seeming less real. But our dreams can be as real, bright, and detailed as wakeful perceptions. I do find that the more rich and vivid a dream is, especially the long vivid lucid dreams, must come by strongly imparted, detailed forceful information, directly from other quantum souls and ethereal aspects. Our created perceptions are *always* created echoes of sensed information from somewhere. So, the source of all conveyed information in my dreams is important to me. And our long, vivid, most real dreams, we call visions, in which we know we are dreaming, are likely communications from the spiritual realm. I find that if a vivid, lucid dream shows me only things of an eternal loving perspective, I know the family of GOD is the source. If it holds dark murky meaning with no eternal discernable truth, I forget it as coming from a confused source with nothing eternal to share. Do not be afraid of dark dreams, no lie can harm you unless you believe it. And if you are one with GOD, evil forces cannot come while you sleep without permission as a teaching trial. And again, the eternal perspective and HOLYSPIRIT are required to know this properly. I know beyond doubt that GOD *never* informs anyone of anything in weak dark or murky dreams. Only the most vivid *lucid* dream clarity and richness, coming from a pure strong eternal perspective, can come from a most powerful good source with purpose. And HOLYSPIRIT is required to properly discern the good or evil message in all such dreams. GOD only sends dreams/visions to those who need them, and they may build our fuller understanding of it all, after sending us a vision. Every choice we all make has its eternal consequences and implications to at least some degree, making each choice a mathematical equation. From the eternal perspective, every choice for this world and self, over eternity with GOD is mathematically a ridiculous. But a temporary materialist perspective, sees the reverse. This is a choice of desire having mathematical outcomes. So, to what degree, does each soul's desired perspective, affect and make their mathematical equations subjective? No one can approach a mathematical problem, from any angle outside their own perspective. Can you show proof that your math's foundations, parameters, and perimeters, are not wrong from my perspective? No. The eternal perspective is required to properly understand every created aspect of wakeful or sleep reality that exists. *Nothing*, not the first thing, not one plus one can be completely known through individual temporary perspectives. What huge percent of the souls in this world do you think sense life, through a temporary perspective? What huge percent do not know what temporary perspective even means? What percent would you say have never once held the eternal perspective? What percent move back and forth seeing just its periphery? And lastly, what percent would you say live only and always in the eternal perspective? It is only FATHER, SON, HOLYSPIRIT, the Seven Archangels, the Four Seraphim, the remaining two thirds of the Cherubim, any lower orders of remaining Angels, and all the Saints of mankind, who live within it faithfully. And the word *faithful* means eternally without fail. I on occasion stumble into the periphery of a temporary perspective; but I sense this and direct my efforts to recover. I never stumble as deep, or for as long, as I once did. Such stumbles are called backsliding by GOD. Again, this is simple to understand, in that instead of making progress, one has slipped back and lost a certain small portion or aspect of their desired perspective. Now the dangers in each backsliding are different. Think of walking along a high very narrow rising mountain ridge. Sensing your way for a sure foot-hold; but then you are distracted and step to close to the edge, and a weakness in your belief gives way. You then slip and slide down the slope, but how far and fast. Some backsliding is easily and quickly recovered from, yet some are fatal. As one can fall near to their beginning, and their faith suffers a fatal injury. My worst slide was the mid-1970's, when my mom's second husband died in a plane crash, and I went to stay with family to help my mom. It took many months to spiritually climb back up to my former point. Never underestimate the importance of any tiny eternal thing, and with growth and maturing of your eternal understanding, you sense more, and stumble far less deeply, far less often, and for far shorter durations. We also sense a correlation, that every time we slip, it is by stumbling into some old, or new unrecognized aspect of a temporary perspective. Now fallen science lies, and dies, by focusing on physical causes, effects, and correlations. But GOD'S science brings all conscious, spiritual, *and* physical aspects together. Please see that in every station of life's awareness and existence, and at every level of influence, from a lowly laborer, to a president, to FATHER GOD above it all, there exists in us all an aspect of self-interest. A leader can seek the greater good for all they rule over, yet also have their desires and life to attend. My point is, those who say the rich and powerful are evil, and care only for themselves, speak and think from ignorance and envy; no one is pure good or evil in this fallen world. Every station in life, influences someone's existence. FATHER above, loves and considers every need and hope for all his children. But being loving means two things: FATHER needs to eternally love, and he needs to be eternally loved. Never judge others for unwitnessed acts, for which you have unreliable, hersay, arguable, or insufficient evidence; be very careful. Now in having studied the current and historical aspects of each form of governance that exists upon the face of the earth. I find corruption and sin exist from the lowest peasants to the highest rulers in each and every form. In this fallen world, the outcomes of most big changes, has been to replace the imperfect, with something even more imperfect. We find the combination of Biblical Judeo Christianity, the U.S. Constitution, and the broad freedoms of Private Market Capitalism in the 18th 19th and 20th century, provided the most just form of modern Government, found this side of Godly rule. But in the 20th century, the U.S. became infected with liberal globalism and Hedonism. Can we temporarily recover or improve? What a monumental question, and what a complex task that would be. Now we can judge committed sin, but we must be diligent to exclude all false, unverifiable, and ambiguous evidence, and never pass eternal judgment. To see FATHER'S new coming created reality, requires the eternal perspective. While all the miracles, suppositions, and unfounded leaps of faith, found in mankind's modeled physical beliefs, are all dead or dying. The paradoxes and problems in mankind's model, appear solely by their beliefs and understandings that created them. Their many problems do not arise in GOD'S true model; so, they are undeniable evidence of fatal errors. *We* see a spectrum of perspective in all souls, from the highest eternal perspective, down to the lowest temporary perspective. We can also see the spiritual traffic going in both directions. Now predictably, this traffic near the two pure perspectives, tends to mostly go one direction, while traffic at the uncommitted center, tends to go back and forth the most. But in each instance, what else could we expect from free-will Dualism. It is the evidentiary results of there being *only* two eternal conscious choices, or directions; and only two final states of spiritual outcome engrained throughout a creation *designed* to find FATHER'S loving family. Also, not so strangely, all souls nearing a pure eternal perspective, see all things more and more clearly; while all who near a permanent temporary perspective, see less and less of any true representation. The latter have past the precipice of blinding themselves beyond hope of escape. These desire to never know GOD'S truth, and this desire they shall have. Now if a soul carefully reads, and says *this* book is all superstitious ignorance, that is strong evidence they are deep in a lost temporary perspective. But the more you see a probability, and then the certainty of GOD'S truth, the more secure you are. When a materialist firmly denies their own forceful soul, there remains no eternal construct for them to be held by. So, they are set adrift in their individual conscious abyss of hell and death. Every truth is repeatedly and totally rejected and discarded by these, over and over. We see all unsettled desires *for*, or *against* GOD, grow and wain toward these two opposing perspectives. Now many lost souls, in their last years, hedge a bit on their assertive denials of the soul; but why? Now at this point, I would like to define what we see as the number value of each sin. By this we mean that if two people are totally alone, and one lies to the other and it goes no further, the value, or destruction of one certain sin has occurred. But if a soul is in front of a group, or on TV with millions of souls watching, and they lie. All believed destruction, and the destruction that is then spread by them, falls at the original liar's feet. Yes, if I or anyone writes a book, we are accountable to GOD for *every* word read by another. I fear the eternal conscious anguish that would come upon me for misleading you and the many; if I were not sure, I would not write. The eternal perspective and understanding says we can never deceive another, without deceiving ourself in an equal to far stronger way. Every confusion, destruction, and division, comes through sin, so *every* sin brings the same within the sinner. Those who secondarily hear, believe, and spread a lie, are also guilty of the number of their sin. Be sure of what you say before GOD, lest you deceive another. Whether intended or not, a lies destruction and divisiveness occur. *I* cannot write or speak or live a lie, without being held responsible and accountable. And the sum total number value of all sin in Heaven and Earth, falls upon Satan who lied first.

FATHER calls Satan the father (meaning creator) of *all* lies. So, Satan's number value of sin, has become incalculable. When I see all the godless lies spread by those in journalism, politics, entertainment, and most pulpits, I fear the thought of being in such accountable shoes. The look of shock and fear in their eyes on their day of account, will be wrenching to see, let alone felt first hand. I think most in their realization will turn and cover their ears, like one does at the impending great hurt or death of a crying child before our eyes; there are things we can't bear to see or hear. The accountability for all evil on that day is dreadful; who can bear to witness *all* the evil ever done, or its coming accountability. Is witnessing all accountability good for GOD'S Children? Will this give us greater eternal strength against sin? And on the side of the lost, will it raise their never-ending haunting agonizing self-punishment of forever trying to understand, deny, and forget. In the eternal perspective, we see all things created by GOD have a loving eternal spiritual *and* far less physical purpose. *We* must seek the eternal spiritual purpose in all things. It's impossible for an eternal GOD to have even one temporary purpose, it just cannot be. Even this now *temporary* fallen existence, filled with temporary things, serves only GOD'S eternal purpose. Every temporary purpose exists in and by a temporary fallen individual perspective. Do not in any self-ruling measure, desire the now temporary physical aspects of creation. It is all a foregone aspect, being as if it were never here at all. GOD must abandon all that is defiled beyond hope. The Millennial Reign will be the final moment of *this* creations searching, loving, and trying of the last remaining souls. After that reign, this temporary Earth and Heavens, and their contents, which *fallen* angels and mankind have corrupted with their temporary nature. These all dissolve at the instance in which FATHER forgets his desire for them to exist any longer. And the layers of the New Heaven and New Earth will firmly appear, line upon line, and precept upon precept as FATHER forcefully perceives them to exist before our eyes. Then the forces of pure good and eternal love will at last, reign over FATHER'S found family. The eternal perspective will have secured GOD'S loving eternal family. And all that did not love GOD and their ways, will have found only themselves, forever sealed and cast into a prison whose spherical shell and center are formed of a wrathful fission and immense gravity. I can't see why souls desire to go the wrong way; yet I once headed that way. But how can they not sense as I did, the *growing* lonely separation from GOD that exists between all who are on that path. I always felt it, but when young and lost, I seemed to stagger and be dragged along by the crowd till it grew unbearable. I looked around, wondering what are we all doing, where are we all going. I could not continue in the ways and confusions of hell, that had no eternal loving meaning or purpose, and lacked any true understanding. Even if every other soul I knew and loved continued, I could not continue to live in the endless lies and confusions of a fallen mankind. I felt so temporary, lifeless, and out of place. And I did not want to be there, as I did not belong there. Then GOD sent me a messenger, and I will love and pray for him always. Do I always sense that where I am now, is where I belong? No. When standing among the blind and lost, I sense no spiritual sending or receiving of GOD'S love and hope being channeled through anyone. I sense I do not belong among them. I feel this often, so I excuse myself. Some say this is rude, but it is the *eternal* spiritual sensed understanding and way of eternal life we all need to live. I don't know why most cannot see that to truly love another, means to care greatly for their eternal soul, and be willing to do anything good of GOD that might help that soul. Yet also sense whether that soul wants or will accept such help at that moment. If I cannot give this help, for which I exist, I leave. Most souls refuse spiritual help, they can only interact in a rebellious worldly physical way. I leave because this is not spiritually increasing or rewarding for them or me. Yes, if a soul is physically suffering from exposure, I will give them cover. If any have not eaten for a few days, I will feed and supply them a few days journey. I will give water for the truly thirsty and aid for the truly injured. But I am not part of any charity that only increases a soul's physical abundance and dependance. The physical body is just not as important. The body is a vessel, a vehicle, to get our soul around to make our choices to do GOD'S spiritual work. A body requires a minimal amount of covering, food, and water. But Scripture says; 1st Timothy, 5:8, "*PERVERSE DISPUTINGS OF MEN OF CORRUPT MINDS, AND DESTITUTE OF THE TRUTH, SUPPOSING THAT GAIN IS GODLINESS: FROM SUCH WITHDRAW THYSELF. BUT GODLINESS WITH CONTENTMENT IS GREAT GAIN. FOR WE BROUGHT NOTHING INTO THIS WORLD, AND IT IS CERTAIN WE CAN CARRY NOTHING OUT. AND HAVING FOOD AND RAIMENT LET US BE THEREWITH CONTENT.*" What more-good is there to physically do or have than this. Yes, I have shelter provided by GOD. It is sparsely furnished with a few things that others think I need, and gave to me. Is this far more than I need?

Yes. Now is it a sin to have more than you need? No. But it is a sin of covetous to *desire* and seek more than you need. Is it a sin to deny a physical request from one *not* in true need? If a drug addict asks for money to by heroin, I say no. But this immediately opens up the broader spectrum of yeses and nos. We see that only our growing eternal perspective and communion with HOLYSPIRIT, can give us the right answer in each case. There are those Godly Patriarchs in Scripture, GOD entrusted with wealth. Was this for them, or was it for them to care for many of GOD'S Children. It was always the latter. I would like to touch upon an example that many people deal with, being that of a parent, spouse, sibling, or friend descending into the horrid depths of dementia. Now I relate my own *spiritual* experience, and offer no *medical* or *psychological* advice in this struggle. I offer spiritual advice to be patient, compassionate, and to pray and fast for your strength. My 86-year-old, legally deaf, blind in one eye, mother, had over her last 3-4 years descended into a horrid place. Our family combined could not care for, or keep her and all those around her safe. Her mind had become such, that she lived in a nightmarish confusion, agitation, anger, and ever-growing fear. She had become a danger to herself and others, as fear is a most dangerous emotion. And in her finally being placed in a facility, we were told she was even more combative and refusing food and medications, which she had done with us. But to get to my point, my siblings spoke and worried much about her needs in a physical way; but I did not share their level of worry about those things. I sensed they maybe thought me less caring than they were, as my concerns were about the fast-approaching destination of her soul. Before being placed in a facility, she was difficult to communicate with, and she lived with family members nearby. For a couple of her last years, I would pick her up 2 or 3 times a week to take her for lunch and shopping, and to my house to visit. I took her to all her doctor appointments to help my sister's family, who bore the brunt of her care. About a year and a half before her passing, I found that when we were all alone, and I spoke deeply about GOD, which I always did. She became unusually focused and aware and sad. She would say she never heard life explained like that. She would say she just wanted to sit and try to hold onto what I had said, knowing she would soon forget. Yet, I have no final evidence of her salvation, or mine. Now I am not saying my siblings did not consider her spiritual life, only that they did not speak about this as their greater concern. I know the eternal perspective of Dualism affirms for us, that perceiving and understanding both halves of our existence is needed to be set in our proper strength and belief. But my mother was greatly diminished in her physical interface, and her spiritual understanding. My mother lived her life mostly and fearfully on the physical side, unaware or considerate about how her spiritual side was diminished and askew. At her last, both sides of her existence seemed so diminished, she existed in a fragmented fearful awareness of a bad dream like state. She did not *properly* see, know, or understand reality *before* her dementia, because she did not want to. At her last, she did not understand GOD'S reality because she could not. She *may* have left GOD beyond her reach and desire. At her very end, I saw and loved such a small remnant of a body and shell attached to a soul's confused understanding and belief. So, now her choice may have been finalized and I fear about her judgment. My siblings think she is already in Heaven. Again, I wanted her to eat, shower, and not physically suffer, but that was a small part of my concern. If any can't see why I write about this, well please know that within everything said and done, there are deep layers of meaningful understanding that never originate with us. Meanings and understandings from long before us, reach down as we each build our beliefs and perspectives. These inherent layers of reason and purpose may suddenly appear to us, because they have always been here. GOD'S reason and purpose exist in every spiritual and physical foundation of everything; but only the fewer desire and find these *deep* spiritual founding reasons and purposes. Most only see the emotional, physical, and spiritual surface. And a growing number, see only the physical surface, period. As a child I was always out of physical step and physical understanding with my siblings and peers. Looking back, I was immature as to the physical ways and means of this world, and I feared it all more than I understood it. I did not realize until later, that by my core desires being for GOD, I was drawn to sense, build, and acquire, a different form of knowing. I was developing my perspective of reality, as was every other child. But mine was most often out of line with the consensus view. I always felt detached and out of place in this world, and now I understand why; and I cherish this sense. I sensed, desired, and was attracted to something else in an inherent intrinsic way, yet, I did not understand this in any way. I now know, that I simply loved and feared GOD above all I saw around me. I am not saying I was a good child by any means. I desired many of the things that others wanted; but I wanted them to exist in a different way, and for a different reason. I always wanted all worldly ways and things to be different than they were. I did not know how, or in what ways they should be different, I just sensed nothing was right in this world for me. I felt tolerated by most, and slightly loved by a few; but I felt known and understood by no one. I know all children can feel this way to differing degrees; but this can be to very different degrees, and for very different reasons. And again, I did not understand this either. Though I tried to exist in the world's temporary ways for a while, I did not then, and do not now, rationally understand temporary desires, or the ungodly choices made by all souls, including myself. If I see anything physical, which seems to raise a desire in me, I ask myself why I would desire that temporary aspect and its affects that fade, but I find no eternal answer. So, if I find the world to be most odd, can this world do less than find me most odd? I think not. I see the world as a sinister place, formed and filled with endless delusions and illusions. Do many in this world see me in this way? I hope not. I hope in peace and love that all souls would find and accept the eternal perspective even more than I. Do those of the temporary perspective, wish I would adopt the temporary perspective? No. Most do not know these two perspectives exist. The true scope of all reality is so much greater than the sliver of the physical universe, fallen mankind's science, can sense. See all reality as two spheres occupying the same space, to represent its conscious and physical sides. As to the physical side, a 6% sliver of it all, along the middle, is all the limited slow small fallen mind and body of mankind can now sense, accept, and comprehend. And most deny the conscious half of reality, believing souls do not exist, or create reality. Though souls of a temporary physical perspective deny these things, these truths taunt and haunt them from the fringes of their quantum conscious periphery; and again, this is the last thing they want to find. This entire understanding was there at the very beginning before all lesser gifted souls; but in most, this truth has drifted into their periphery of lost desires, being now a rare fleeting flash in the corner of their eye. The truth is still there, but not enough to be sensed through their current fallen perspective. They have become, as said by the band Pink Floyd "Comfortably Numb" to this all. That song may be this fallen world's true anthem. Can we all accurately and honestly know and describe our own perspective? Is your perspective individual, physical, and temporary? Can you discern the temporary or eternal perspective of our family members, friends, or colleagues? Can you sense how mixed, strong, or pure, their perspectives are? By HOLYSPIRIT'S gift of discernment, we sense such things. But do you truly want to sense and know these aspects about yourself and others? To want this, is a vital inherent attribute of the eternal perspective. The eternal perspective desires and endeavors to see the forces in every moment of everything, and everyone through all true eternal means. The eternal sense, sees how all things move for, or against oneness with GOD. And we know the remnant of this physical reality is soon to be lost, meaning it is all temporary. But it is a lesser part of a much larger wondrous eternal aether. The eternal perspective sees life from before and beyond all the temporary fading aspects of reality. To know GOD'S reality, our eternal perspective is our most precious required sense. Consider, that to physically grasp and comprehend the size of the earth, this is best done from out in space; not standing on the ground in the middle of a dense forest, reducing a fallen soul's perspective to their close immediate surrounding. So, to understand reality, also requires seeing it from outside of it all with FATHER, and never by studying its tiny pieces from deep within. One must also accept all the greater *meaning* of reality, that only an eternal vantage point reveals. Become one with the mind of GOD, and step out and beyond with them, and embrace the true eternal scope of reality. You have nothing of love to lose, and everything of eternal love to gain. Consider the beauty, form, and indescribable complexities and fine-tuning we see in this universe, and all life. Consider your own awareness of it all from the eternal perspective. We see it is impossible for all entangled beauty and complexity to come to exist by random physical chaos and disorder after an explosion of a miraculous singularity. It is impossible, for a mathematical illusion to inflate nothing, or next to nothing, into endless forms of everything. There is the saneness and insaneness of math. There is a point where math, reality, and the mind, do *not* work without GOD. To say that anything that might mathematically work on paper, can physically work in reality is insane. I hope you allow yourself to see the incalculable odds of chance, in forming the great complexities of life in one cell, are unimaginable, impossible, and also insane. You could write down the chance of one, in a stream of nines (#8 font) a million miles long. And never come near stating the odds of life and this reality, starting from a physical quantum fluctuation, in a once pure energy singularity. Again, we stare at their profound founding miracle of all combined energy and force, being unimaginably compressed, condensed, and sitting in the middle of endless nothing, with not one single awareness of it. Their faith in the countless miraculous events this requires, comes not by any sane logic; but by their core desires for it to please, please, please, be so. It sounds like a ridiculous fairy-tale, atheist tell their children at bedtime. Why would all that is, this vast universe, be compressed into an indescribably small singular point of pure stability, sitting in the middle of endless nothing. Waiting on an unfounded unexplainable interior fluctuation to appear in its perfection. For any true pure singularity to exist, it must be a perfect stable single form and quality within. So, from where does their faith and belief in a required imperfect fluctuation come from. A Materialist belief, is no more than an endless stream of impossibilities. They believe if they just keep their mind closed to GOD, and keep answering their own consciously designed physical riddles as they go along, that they are moving towards something real. But *we* know conscious patterned forceful creative guidance is a necessity, because we are all consciously here, desiring to exist, and we all forcefully project, perceive, and create our substantive perceptions out of quantum waves and particles. For many believers, this form of forceful creative soul could not exist on a vastly larger more powerful scale than us. For many, though they see every other reality existing in its large to smaller forms, for them, this cannot mean this occurs with all creative souls. Is FATHER, being on a far greater scale than us, far more aware than we are? Did the three greatest souls in perfect agreement, forcefully/substantively perceive the vast beauty and complexity we now sense secondarily? This is what is sensed and found beyond reconsideration through the eternal perspective. FATHER is with zero doubt, shown to me, to exist *necessarily*. And by him alone, we are eternal gifted souls of awareness, who like our FATHER (meaning creator) forcefully create and know our personal existence within ourselves. Yet we, in eternal need, reside within the great external reality created by FATHER. And together, we form all free-will loving reality; but why do most not see this? By our core desires, we choose to form our beliefs from the possibilities and probabilities we each desire and focus upon. If you stare only at a lie, it is all you will see. And if you stare at the truth, and a lie, you may still be confused. But if you stare only at the truth, you will never be lost. The only way to see the vast complexity and intricacies of life existing now in their evident increasing disorder; is to see disorder was introduced, only after the perfect creation of life. And by our *creative* souls existing now, in a universe that all lesser souls combined could not, did not create. The odds of GOD existing, by any honest scientific tally, are a near certainty. If one desires, one can reverse engineer the universe's stretching in a physical theoretical way, relying on the slowest particle of light, and theories built upon many physical miracles, and see a very old universe. By mounting problems and failures in the Big Bang faith, I see only its impossibilities. What I am called to write, deals with all the foundational spiritual and physical aspects of reality. And in *this* second book, GOD pushes my focus out into the spiritual periphery where our *eternal perspective* is sensed to grow. And I myself, find it spiritually edifying. I don't think I have shared before, that I do *not* see, ponder, learn, *and then* write something, only after I have come to fully understand it. Very often, I see and write, and then later learn and come to better understand what was written. I write in faith; not always, but often. And as I learn, I wonder if anyone else desires, hears, and comes to know these things. Surely, I cannot be alone in this. Again, not one word of this understanding is my own. I claim no perspective as my own; all I share is not mine. This is the eternal shared perspective of GOD. I only know these visualized understandings from above, have passed through my soul, and hopefully yours to be passed on again. I am so grateful for this gift of understanding as I search for others who hear. What greater joy can there be than souls becoming one in the eternal love of GOD. Confusion is spread by lost science, as they physically interpret and imply their desires into all sensed information, being quite uninterested in what was initially intended, as GOD forcefully projected that same once perfect information. The one perfect eternal omniscient interpretation is vital. Some say, who do I think I am, that I should be given GOD'S truth, and not them? And I also ask who am I, that GOD should reveal to me his wonders? But GOD showed me that he gives this to all who truly want it, I am in no way singled out by GOD. So, the true question is, why do many refuse it? I bow and put my face to the ground, and inch my way in fear towards our LORD'S feet, from a far-away distance. And as I fearfully want to inch closer, I ask GOD, who am I, that I should be allowed to move an inch closer. But GOD shows me in their ways, I am loved, as they call me closer still. Again, visualize yourself sitting at a table and seeing the entirety of reality before you. The firm surface of the table, representing this world's temporary perspective of sensing only a sliver of the firm *intermediate* aspects of the FATHER'S external creation. See your clenched fist high above the table's surface, representing your soul's tight hold on all aspects and elements of the eternal loving omniscient perspective of GOD, ruling over the physical. And knowing GOD intimately, you are held in all his understanding to be separate and above all the temporary physical aspects and elements below. Now why does GOD always use the word fell or fall when talking about mankind after their rebellion. See your fist held high opening and letting go of GOD'S truth and ways, and falling into an open flat palm resting on, and becoming one with the surface of the physical table. This is the literal fall, or felling GOD speaks about. And from this visual, what would be GOD'S redemptive plan for our fallen soul. Is it to leave us there, and endlessly forgive us for being a physical creature, stuck in destructive divisive rebellion. Or is GOD'S purpose to raise and restore us on high, to become the perfect eternal sons and daughters we were created in all hope, meaning, and purpose, to be? Are you strong, are you a clenched fist holding GOD'S truth and ways, high up in the eternal perspective; or a weak flat confused open hand, being one with the physical surface of reality, still lost in death and hell? We must recognize that without loving, and receiving first-hand, GOD'S truth, instructions, and guidance; all one can do is lay there flat and physical, and form false belief after belief, according to one's current meandering desires. If you hear gibberish coming from an insane individual, please know they are insane, because they believe insane things; and there are endless deviations to insanity. How are you assured that the world at large is not on some level, filled with insane beliefs? What is it that makes a belief insane? Simple; if it is not true, it is insane. Insanity means separated from the truth of reality. An individual can easily create a personal delusion of religious expectation, at the death of their physical body. But GOD does not provide each souls desired expectation of an individual heaven, so these receive nothing. Why should FATHER eternally work to provide joy for all souls he has gifted, without his needed eternal love in return? FATHER has worked to perceive and perpetuate so much complexity and chance and choice. Do you see that before the beginning of all we see and know, that FATHER deeply considered all things. And understood *before* gifting any eternal souls, that free-will eternal love may not come. But in his first gift of his SON, eternal love was found, and FATHER proceeded with his plan. A great *eternal* never-ending responsible work is no small sacrifice for FATHER to consider and accept. FATHER existed in his lonely zero responsibility for, and over others, before he gifted any souls. FATHER accepted eternal responsibility, effort, suffering, and loss, in order to find eternal love. Can you understand such a vast eternal commitment to find and have eternal love? Are you truly thankful for FATHER'S never-ending work to nurture and perpetuate your love? FATHER'S work and MESSIAH'S suffering Sacrifice and the gifts of HOLYSPIRIT, show us how much GOD wants, works for, and needs, our love. And our returned true eternal love, is shown in our never adding to their great work and suffering, beyond their already great required measures. GOD must, and has chosen to bear this work for us all forever; but they derive no pleasure from the pain of wasted work, spent to find and have that love. Never forget this, as it is part of our shared eternal perspective. Now it is often said as an analogy, we must peel the layers of an onion to get to the real truth. Of course, all we find at the center of a real onion is more of the exact same onion. So, is GOD'S truth found at the core of what each individual chooses to peel? Or do they just find their own created belief? In a spiritual and honest peeling away of the layers of reality, down past the physical facade to the core, past the layers of what most want reality to be. We find there, at the conscious spiritual center of reality, the polar opposite desires we see in the perspective on mankind's surface facade. I know all can simply reply to me, that is then, just what I desire to believe at my core, and ask how is it superior. And where is the proof that what I desire to believe is true or superior to what other souls desires to believe. But this is my continuous point, that by my core focused passioned desire to surrender my errant hopes, to GOD'S every written word. I surrendered all *my* individual ways and desired beliefs in a dark field, on my knees in the spring of 1974, asking to receive and accept only the truth interpreted by HOLYSPIRIT. Either we overcome our individual physical sided beliefs, or they overcome us; and with every word, I express only this. Now, excuse my use here of a card game analogy, but we must first make sure we are all playing with the *one* proper full deck. Compare the many differing decks of cards used by this world, against the one true deck. Only then can you look at the hand I play, dealt from above, which by my free-will choices, I am drawing to. Then truly see the hand you hold and are drawing to by the free-willed choices you make. I *do not* imply, so do not infer by my card game analogy, that by the physical elements of disorder and chance introduced by sinful imperfections, that finding GOD is in any way, a matter of the luck of the draw. It is a matter of loving choices. Good choices, means holding onto cards of eternal understanding, and discarding all cards of individual understanding. We are drawing and discarding cards with each new choice we make. And as long as we always draw from the *eternal* deck, and hold only eternal cards, we are living by the redemptive and immutable promise of GOD. I wish a life-long card game was not an analogy of life's choices, because card games *are* associated with luck and gambling. So, see this analogy as a card game of eternal loving strategy, void of luck or gambling. And where everyone must first choose either the one eternal deck, being dealt by HOLYSPIRIT, or one of endless temporary decks. And then our choices to win or lose begin to appear as each soul's hand must be played out in free-will desire. In our choosing the right deck of GOD, and holding only the right cards of the eternal perspective, every hand can be played out to win, no one can be dealt an unwinnable hand. But by gambling with a temporary deck, no one wins; all temporary decks are set as unwinnable. Each soul may learn the rules and play the eternal game; they can also try to cheat, or falsely accuse another of cheating, but in this they can never win. We see so many temporary games and temporary ways to play one's hand. But in so doing, we see only countless ways to lose. We see many *individual* perspectives, but only one shared eternal Godly perspective. People complain about the perspective of others all the time. People are always saying, if you could see this or that through my eyes, or from my perspective. Yet so very few say, if we could all see reality through GOD'S eternal perspective.

Most souls go their entire life, never hearing, thinking, or caring, about what is the eternal perspective. Even in most places of religious belief that fill this world, only the few know of this. But why don't they all know? Yes, most preachers teach from a temporary physical perspective, making them false ministers of light. Once we truly see the temporary perspective and its ways of power, privilege, and possessions, we can see how we have already lost everything physical we have or desire in this world, before we ever received it. It's actual loss just has not come to our awareness yet. But do not doubt, your awareness of complete physical loss is coming. From this world's fall, it all became a vapor, a fading echo, a blurred reflection, a dimming last twinkle of light, a wish that can never come true. This world fell in an instant of GOD'S recognition, and became destined, as if it never were. If it were still of any eternal value, it would not cease to exist at the final separation. GOD sees it is foregone in its diminished semblance of ever-increasing imperfect defilements. GOD turns his eyes and desires away from this world more and more. Do you feel in this moment that you are temporary; or do you see the truth that you are an eternal soul? If you firmly see this, you have an eternal desire at this moment. If you sense reality must be viewed differently, then you have begun *your* eternal work within your soul. Take CHRIST'S hand right now, and receive HOLYSPIRIT'S teaching, strength, and comfort, and seek and acquire the loving eternal mind of CHRIST. There is only loving gain, and all hateful loss. I love and know GOD above, and seek all his children of love and hope, waiting here below. Most all souls below are now contrary and adversarial to GOD. But out of it all, love and hope spring up here and there. Some grow for only a while, but others grow eternal. Please see that without intimately seeking out, knowing, and becoming one with GOD in their eternal ways, to which we are accountable as one family, there can be no truc growth in you at all; for what else would you grow towards and into? What can lead you, if not the HOLYSPIRIT'S eternal perspective. There is nothing else left but fallen mankind's confused changing desires towards GOD, family, friends, their selves, and this world of empty temporary things. Look around, try to find wisdom in this world to eternally lead you; there is none. What musical chords of love does GOD strike in us? Do they strike our emotional keys, spiritual keys, and logical keys, or revelational keys of a loving reality. GOD must play them all in us, for an eternal melody cannot be played or heard in any lesser way. GOD only speaks and works in all fullness, using every aspect of truth found in their eternal ways of reality. GOD never tries to reach a soul, in weak or confusing ways. Reality must come into true focus, it must be seen and known as being FATHER'S consciously intended, projected, and entangled creation, even as we focus on any of its lesser aspects. We *best* see another's eyes, seeing their whole face. Seeing how the eyes belong to, and relate to the expression of the face, and not seeing the eyes alone. It is one creation, and every relative aspect and element are to be perceived in one whole and eternal way. GOD, please show us where *you* are and go, and all *you* do, that we may learn to follow you. GOD teach us your eternal love, so we may find you high above. Simple children's rhymes; but JESUS said we must become as little children before FATHER. I view myself before GOD as no more than a five-year-old learning child, and FATHER GOD is my only FATHER. On earth I had a mortal dad, but he was never my FATHER, meaning Creator. Like CHRIST, I long for FATHER'S smiling approving gaze, I trust his every word. Like CHRIST, I know FATHER will never hurt me, or lie to me; he will only love and take care of me, and teach and guide me in my inability to do or know anything eternal by myself. When we see a loved and loving trusting child, walking besides their dad, smiling and trying to match their dad's great stride. That is who I want to always be to FATHER above. I hold no desires to be a GOD above any; I would be the worst at that, being unimaginably inadequate. But since we have a perfect loving omniscient FATHER already, it is perfect for me the way it is. We have our adult responsibilities over family and friends here. But if you remember being five or so years old, not a care in the world, feeling safe, secure, and provided for; is it possible to hold to both? Can we have adult loving responsibilities for our children, family, and friends; yet turn and feel like a child, looking up to FATHER? Yes. It is FATHER'S perfect top-down design of all things. Some think I am a Gnostic; I am not. Like all lesser worldly philosophies, Gnostics have some valid and invalid points. Now as to my earthly physical dad and mom, I honor them not from any temporary worldly perspective, but from an eternal spiritual perspective. And these are quite different in love and hope. I truly wish with all my heart that GOD would tell me all that I should do, step by step. Then I might do things pure and good like JESUS, and I would know FATHER is pleased in his love for me. John 8:28-29, "*THEN SAID JESUS UNTO THEM, WHEN YE HAVE LIFTED UP THE SON OF MAN, THEN SHALL YE KNOW THAT I AM HE, AND THAT <u>I DO NOTHING OF</u>*

MYSELF; BUT AS MY FATHER HATH TAUGHT ME, I SPEAK THESE THINGS. AND HE THAT SENT ME IS WITH ME: THE FATHER HATH NOT LEFT ME ALONE; FOR I DO ALWAYS THOSE THINGS THAT PLEASE HIM." By HOLYSPIRIT'S interpretation of reading the Bible many times, all Scriptures become perfectly congealed in a relative entangled meaning. This is a great proof to me, that Scripture is not just GOD'S inerrant truth, but many layers of the same. I first read and heard its tens, then its hundreds, of perfectly laced truths and understandings, and now in spiritual growth, I hear its thousands; yet it never alters its mooring or course. It enhances, yet never changes one small meaning, it only grows in its perfect strength and harmony. I know with all my essence that mankind could never write a book that could begin to do such a thing. After 40+ years of reading it, it still grows in strength, and I feel such sorrow for all who interpret Scripture for themselves, for they can never hear its truth. Now, it is said by those of many literary persuasions, that one writes about what one knows, and one only knows what one lives. So, writers write about life's experience, through their perspective of it. But I write about my learned perspective from another, and about CHRIST'S belief being his learned perspective from FATHER. It has been said by many that “we see the world, not as it is, but as *we* are”. This quote has been traced back almost 200 years, without a proven initial source. Throughout history, souls have said life is a spiritual journey, not a physical one. Now the truth was long before all false modern science. And life itself, cannot be described in any physical way that conveys the truth about us, and or reality. Do you hear me share my eternal journey, that began at my spiritual birth in 1974, to where I spiritually am now, and on to where I must spiritually go? Or do you hear the philosophical, theological ramblings of a physical creature? I hope you hear so much more than that. Every thought and choice we all make from our gifted creation, has been our spiritual journey to bring us to who we spiritually are in how we relate to and desire GOD’S eternal plan; and this is the standard by which our spirit is eternally judged. And upon our spiritual resurrection in our *true* repentant water burial Baptism, the old person dies, vanishes, and never existed, all according to CHRIST. I *must,* and firmly *do* accept this from my LORD as being the truth of my reality. I see the physical reality in going forward as my periphery, and platform, and backdrop, to the spiritual world. And I see my forgiven past in this same way; it is a vapor that no longer has an eternal meaning. Our environment is not our reality, it is a stage that allows what truly matters to be played out. And that stage must remain perfect as created, or its defiled and must be replaced. The eternal perspective builds nothing less than our perfect understanding. We must be a spiritual creature, see the spiritual, think spiritually, hear, speak, and write, spiritually. We must know, remember, and be, all things spiritual. If anyone sees, thinks, hears, speaks, and writes, from any past, current, or future, physical perspective, they are a physical creature, and not a spiritual creature. The physical body our soul moves through, is a lost fading aspect in our journey. The relevant aspects are GOD'S beliefs, their loving reasons, purpose, and the outcome of each child's spiritual journey; these are the eternal aspects. Don't focus on the physical path, the fallen physical reality has become an immaterial aspect. The *sum* of our eternal perspective is not a small, thing that just is, or that can be suddenly added. It is our foundational reordering of our relativity, reason, and purpose; it is our new spiritual sense of emotion, sound, sight, and desire. It is our sensing the conscious and physical worlds in completely separate ways. The step-by-step growth process *at first*, will not feel like I describe it; but it can grow to be. For most, one should *begin* to feel this way within six months to a year or so, and get stronger with all adaptation. I had never heard of the eternal or temporary perspectives; so, it took much longer for HOLYSPIRIT to show me. It was a spiritual path that took *me* decades of accepting small pieces in faith before finally seeing the picture GOD was revealing. It was an arduous work *for* GOD to pains-takingly build my true understanding to the point that I could see this congeal. But I hope to quicken the journey for all who embark upon it. As my confused understanding of a temporary perspective faded, I saw my eternal perspective reinterpret my memories. We can't remember things in the former way that no longer makes sense to us. We always remember things through our current over-all understanding. So, in drastically altering our understanding, our memories take on new meaning. This has a spiritual feel, like a form of memory healing. The world may call me crazy, and I care not. I know I have never been more-sane in my life. I see all things congeal and move in one rising direction of growing consistency. My knowledge, reason, understanding, purpose, emotions, and desires, are all aligning in a way I never experienced before. My youthful illusions of reality were very fragmented in what I believed through a temporary perspective; which in turn, fractured my soul in those same unrecognized ways. I had no true eternal point to align all my aspects towards as a common purpose and spiritual destination. But now my road leads to GOD. I can now communicate spiritual things, in far more relative concise ways. Yet I find only fewer who desire to know very much of this. A strong final conscious divide in reality is forming; which is not in any single way, an arbitrary act of GOD. It is a divide, created by all desired free-will itself. Which GOD must soon seal off to protect the eternal spiritual love and peace found on the side with him. And the gifted souls, being parts of FATHER that have left him, will be eternally lost to FATHER as his great sacrifice; protecting us from that never-ending free-will hate and anguish on the other side. FATHER created the means; and shares all his understanding and reasoned purpose for *allowing* this divide to occur. But the divide itself comes by choice in all souls. I have searched the reaching ruins of mankind's spirits, and searched the ruins of the physical creation, yet I have *never* found one point of division, that originated from FATHER after his perfect creation. GOD does currently see, chastise, and try to get his children to correct our every divisive aspect, and commands us to keep our soul *and* body sinless. It being a temple for HOLYSPIRIT and our soul together. GOD tries to help repair all spiritual divisions, but at the last, FATHER must seal off all reprobates. GOD wastes no efforts on souls that have consciously journeyed beyond possible correction; what would be the eternal purpose. Does every soul sense crossing this line of permanent confusion, crossing from the side of possible correction to the other? Can a soul ever find the truth again, after leaving every last hope of knowing GOD'S truth behind? No, to both. But if any hears enough to turn now, this is *their* evidence of being on the reachable side. They still hear, and might live the true purpose and reason for life. No one can turn and go towards a new direction they cannot sense or desire. If you hear any of this, please know you have at minimum, turned to look. So, run to GOD that you may be with all who truly love you; but only if knowing and loving FATHER and his Family is your core desire. And without doubt, GOD and his Family want to eternally know and love you. We all exist in the unfolding of all that will come; but if you and I should end up on opposite sides, and look across to see the other, we can never know the thoughts or feelings the other has towards us. But they will be contrary as one will be confused and anguished, and one, sad for the other. This world is now an imperfect confused place, and the heavens are also infected outside the Holy of Holies. Physical life is now a most imperfect decaying thing, as confused souls still choose death. An eternal conscious based reality is something most souls do not want; for them, seeing both halves of reality entangled in consonance, is set beyond their temporary perspective. On their own, mankind could never begin to unravel GOD'S mystery. We *can* turn to hear and follow GOD'S truth; we just can never find it on our own, being set beyond our reach. Ecclesiastes 3:9-11, "*WHAT PROFIT HATH HE THAT WORKETH IN THAT WHEREIN HE LABOURETH? I HAVE SEEN THE TRAVAIL, WHICH GOD HATH GIVEN TO THE SONS OF MEN TO BE EXERCISED IN IT. HE HATH MADE EVERY THING BEAUTIFUL IN HIS TIME: ALSO HE HATH SET THE WORLD IN THEIR HEART, SO THAT NO MAN CAN FIND OUT THE WORK THAT GOD MAKETH FROM THE BEGINNING TO THE END*." *No man* can find this on his own, yet GOD reveals these things to *their* Children. Now let's examine this aspect called the wave function *collapse*. From the eternal perspective we see this is the process and moment where a soul's every intentional projected forceful desire, aligns the rising particle layers of each soul's complex perceptions. Meaning, from our creative ability, we project forceful patterns of desirous intent, that bind and align our complex quantum structural perceptions out of the aether's singularities. To be built large enough that fallen mankind can sense and comprehend them. Now the rising unseen wave itself is a sea of congealing unimaginably small points, being part of that rising creative process. When materialist say that physical reality is lost at a certain point of reduction, this is because they can't see, accept, or comprehend this sea of minute lesser particles and singularities that are there. A reduction process goes much further down than mankind can ever know; down to the single pure physical singularities that fill and form FATHER'S spherical aether of reality. Through the eternal perspective, we find complete understanding, minus all the gaps, unseen variables, and paradoxes that plague mankind's model. Like Schrödinger's Cat, paradoxes appear in their mind, when they refuse to see FATHER'S immense wave functions, collapsing to form the universe. If I open an imagined box *in my mind* to find a cat. It is I, who collapsed that wave function, and perceived a cat to only *then* be dead or alive; and it could be either until I do so. But I can have no affect on a cat that is part of FATHER'S ongoing external creation, that we perceive secondarily, and being in an already collapsed state, inside a box. FATHER perceived this dynamic world, filled with his complex process of life, about 12,000 years ago, and there is no paradox. A limited perspective purely physical model, filled with unknowns and problems, has no credibility to me. Day by day, mankind's need for a new model increases within all sciences, because it can never get from point A to Z, it barely gets from A to B, and it can never fulfill its promises. Their theories come by surmising that requires desiring eyes squint at their data in an unease that permeates all science. The old Materialist guard, refuses to concede defeat, and clutch tightly to beliefs having no existential value, and lies that do not exist beyond their belief in them. They indoctrinate the next spiritually blind, physically entrenched youth, to take up their lost battle. They build a desired hollow false continuum of non-reality, and refuse to let go of the weight that drags them down into hell's abyss of endlessly changing illogical conclusions. But *we* place an eternal value on the created ways and love of GOD that allow all realities to exist before our eternal awareness. We are all gifted and created with our Godlike creative awareness of life, for just this purpose. It is a most wondrous eternal gift, and such a tragic choice to set one's soul upon a course of minimizing life and love forever. Biblical wisdom says that at the last, around 300 million souls; which includes 200 million remaining Holy Angels, will go with GOD into life eternal. This means around 100 million out of a rough estimate of 70 billion souls born after Adam and Eve, are saved from among mankind. Being around 1 out of every 700 souls. We know the cost of eternal love is great; and by this number we see how fearfully great. The original dynamic reality that was once created perfect, was created by the one mind who desired it all. To deny this, is to say life came of no eternal purpose, as an impossible meaningless physical incident. Most see love is fragile, rare, and dying, all around them. All souls value their love, but the difference is in how much, and how their love is defined. To value GOD'S love, is to protect and preserve that love eternally. Who can love eternally? A temporary perspective only loves in temporary way; as their efforts and ways to protect and preserve love are so insufficient; this is the dying tragedy of a temporary perspective. Now as we build our eternal perspective, we use the terms Godly love, and eternal love, and innately know something about the giving and receiving aspects of this needful desire called love. But we must learn to see this wondrous mystery in the way GOD sees this mystery called love. Let's start with the fact that we all know love comes in varying degrees. We say we are *falling* in love; we say I love you more now than before; or we thought we loved someone, but it faded. This shows us this world has only measures and kinds of diminishing dying love. Many believe and feel a first true love towards GOD, and GOD recognizes this love; but this love does not always grow or endure. We see souls fall in and out of love, even with GOD. So, what kind and measure of love endures for eternity, *or* what kind and measure of love can then be enshrined by GOD in a perfect environment? After all, that is what the physical creation of Heaven is for. Heaven is a perfect physical sanctuary to seal, protect and secure the eternal love of the family of GOD. We see that love must freely begin, and must freely grow; but must a very *great* true love, once found, freely endure for eternity? Or can it be sealed, cherished, and nurtured by GOD to eternally enshrine such a wondrous treasure? Yes. But only in a created Heaven, by FATHER who has the desired understanding, force, and eternal need to do so. Many souls are falsely taught to believe that their first initial love towards GOD is their final loving step. Many are falsely taught that by an initial loving prayer for forgiveness, they are sealed and destined for Heaven no matter what they do or become; this is not Scriptural. We each can freely of our own desires love GOD; and we must freely of our own desire, grow this to become and remain our greatest love. Once our love for GOD *does* rise and reach up to literally be our greatest love; GOD can then intervene to cherish and preserve that which he has so long and hard labored to find. That which GOD has suffered and endured so much sacrifice, pain, and cost, to have. So never say love cannot last forever. Once love is real, GOD can erase and remove all created things and memories that come against love. Please see the loving relationship that souls can have with GOD, as being beyond all other loving relationships. So many say they love GOD; but what is their understanding of intimate love for GOD? Will your love rise to be eternally sustained under GOD? This you can only know after GOD reveals all things. I know I am called by my love for GOD to sin no more, and be filled by HOLYSPIRIT. I'm called to study, know, and live *now* in the eternal loving ways of GOD. I am informed and warned throughout the Bible, that only by feeding and growing my love for GOD, can I, or anyone ever hope to find these things. I am called by my love to want and strive to be perfect before GOD, and our perfection comes in emulating CHRIST'S love and life. We know CHRIST can forgive a lifetime of sins, in one act of mercy. But how many times can JESUS forgive us if we continue to sin against his love. Spiritually loving GOD, is to dwell within that love defined and desired by GOD, not a love born of our own definition, filled with physical desires. What GOD loves, I love; what GOD hates, I hate; the way

GOD lives, we must live; what GOD believes, we must believe. GOD has no desire to eternally sustain endless forms of unfaithful love or belief. Those endless forms of individual love and belief, define this lost world. GOD wants for me to be me, and for you to be you. But we see there surely comes a point or extent in being me or you, where we become a contrary force against GOD'S family, being divisive and destructive to eternal love. Individualized and individuals, are terms we use to define souls who live across that line. Those who in *their* own current ways and beliefs, are divisive to the unity of GOD'S family. It is GOD alone who holds the belief, the understanding, the conscious size and force, and the entire physical half of reality, to sustain us all together. It is a simple thing to understand, yet many complicate it within their souls by false desired beliefs that deny the true and real. These souls lose touch with GOD'S truth, and grow cold, and deny his truth exists. Do you see that in wandering *far* from the true understanding of reality, how it becomes more and more, an improbable journey back? Do you see how political science has perfected the art of antichrist propaganda, to instill mass godless psychosis to divide people from GOD? It has been known for ages, it is *easy* to deceive the young; and then by pride and self-righteousness, it is *hard* to correct or convince them they are deceived. So, evil fills the systems of all education, to indoctrinate and steal the young. Riddle me this; why would liberals, push to teach our too young children, about victimhood and gender dysphoria? They say they only want tolerance, but I promise you beyond all doubt, they are Marxist pedophiles at heart, grooming your children to increase the number of victims and the gender dysphoric. Stealing all children from GOD is their desire. Allow this, and your children will look at you, the way their teachers do. You can deny this, but in that you have read this, you can never say to GOD you were not warned. Dark individuals, lead the individualism consuming reality; and it was foretold, GOD will not stop it until it has run its course and divided all souls. For FATHER'S righteousness, a free-willed true and immutable conscious divide must come on its own. FATHER cannot cause this divide; and only by finding that very last possible eternal love, can he declare that final divide has been found. FATHER alone, will know that dreaded moment, and put an end to all sinful choices. All *lost* souls, will then be set in a way of knowing they have been found wanting, and are forever divided out and away from all that is loving and good. They will be set in a form of self-awareness that ensures they can never enjoy a new act of individual pleasure, ever again.

They will crave all physical pleasures forever; but have only their every regret to ponder. They will want new things forever, but will never again feel the reward they crave and seek. These will forever consider their past in a shell of a body, as their soul is bombarded by all former created evil perceptions. A body that barely moves or feels, under the pull of a great gravity; surrounded by the fiery wrath of GOD, and the lost. Go there if you desire; I see no love or life in that at all. I long and strive to go in the opposite Holy direction; and I eternally love all who strive and travel the same path as family. GOD says, in his eyes there are his Children, and the Children of the Devil. This means there are only two final families. I say yes, love your worldly family, but never as much or more than your eternal family. If you do, and if they do also, you they will one day come to hate each other forever with a passion that defies all logic. There will exist zero love beyond the Kingdom of Heaven. This strongest spiritual truth is only shown to all having the eternal perspective. We could endlessly write about how our every way, and conscious, spiritual, and physical aspect and attribute, align to either the eternal, or temporary perspective. But I hope we've provided enough examples to give you the sense this is true. I am abundantly thankful to GOD for every word, breath, and heart-beat, they share with me. And again, *who* am I, or *who* are you, to hear and understand these many aspects of life, desired, gifted, created, and provided by GOD. If we pray, and our prayer is *never* affirmed or fulfilled, FATHER says we prayed for the temporary ways, things, and desires, of this fallen world, and not GOD'S eternal Kingdom; so, we "*ask amiss*". James 4:1-10, "*FROM WHENCE COME WARS AND FIGHTINGS AMONG YOU? COME THEY NOT HENCE, EVEN OF YOUR LUSTS THAT WAR IN YOUR MEMBERS? YE LUST, AND HAVE NOT: YE KILL, AND DESIRE TO HAVE, AND CANNOT OBTAIN: YE FIGHT AND WAR, YET <u>YE HAVE NOT</u>, <u>BECAUSE YE ASK NOT</u>. YE ASK, AND RECEIVE NOT, <u>BECAUSE YE ASK AMISS,</u> THAT <u>YE MAY CONSUME IT UPON YOUR LUSTS</u>. YE ADULTERERS AND ADULTERESSES, KNOW YE NOT THAT THE FRIENDSHIP OF THE WORLD IS ENMITY WITH GOD? WHOSOEVER THEREFORE WILL BE A FRIEND OF THE WORLD IS THE ENEMY OF GOD. DO <u>YE THINK</u> THAT THE SCRIPTURE SAITH IN VAIN, THE SPIRIT THAT DWELLETH IN US LUSTETH TO ENVY? BUT HE GIVETH MORE GRACE. WHEREFORE HE SAITH, GOD RESISTETH THE PROUD, BUT GIVETH GRACE UNTO THE HUMBLE. SUBMIT YOURSELVES*

THEREFORE TO GOD. RESIST THE DEVIL, AND HE WILL FLEE FROM YOU. DRAW NIGH TO GOD, AND HE WILL DRAW NIGH TO YOU. CLEANSE YOUR HANDS, YE SINNERS; AND PURIFY YOUR HEARTS, YE DOUBLE MINDED. BE AFFLICTED, AND MOURN, AND WEEP: LET YOUR LAUGHTER BE TURNED TO MOURNING, AND YOUR JOY TO HEAVINESS. HUMBLE YOURSELVES IN THE SIGHT OF THE LORD, AND HE SHALL LIFT YOU UP." "*Ye have not, because ye ask not.*" meaning, you have not the eternal things, because you ask not for eternal things. *"You ask amiss"* meaning you ask for temporary things. What do the lost souls of this temporary world argue, strive, and war about among each other. It is physical comforts and pleasures of a temporary life; it is to live their temporary life in all sustainable abundance. And by such foreseeable abundance, these temporary aspects bring fulfillment, and ease their lost minds. They find temporary solace in seeing their desires are met for their foreseeable future. Souls of a temporary perspective only want more of what they temporarily want; and they want it in quality and quantity. They want whatever it is they temporarily want, to be there when they want it. Souls of a temporary perspective want it all, but this world has not near enough to fulfill all souls. So, who is temporarily fulfilled; it is those who strive to get or take. Now seeing most *professed* Christians, hold to a temporary perspective, do they also pray and ask *amiss*? Yes, they most often do. When I first started reading the Bible, I read it over and over as a book, from front to back intently, for up to five or six hours at a time every day for months. From my first time reading it through, I understood certain immutable existential things. I read and learned early on in great fear, that most souls who turn to GOD, are quickly compromised in their effort by their desires for this world and others. Scripture says most try to find GOD using the same physical temporary perspective of life that led them away from GOD. GOD says most souls desire false doctrines (interpretations). It was this, GOD says they saw happen over and over; so early on, being deceived became my strongest fear. And seeing myself unworthy of ever become a perfect Child of GOD, I long ago prayed that at minimum in all my efforts, I would not be among the deceived. I prayed always that GOD would break through to me with *their* truth; that I would know only GOD'S truth. I prayed *so* often in true fear, that if I am to ever be lost, let it be for other core flaws or reasons, and not by believing a lie. I greatly feared being spiritually deceived into a false religious belief. I know now that my prayers turned out to be a true eternal perspective request. They were purely spiritual, and held nothing with a temporary perspective. So, according to my simple honest core request, GOD has answered my prayer, which was *not amiss*. I've come to see that FATHER answers every pure eternal perspective prayer; it is who he is, he cannot deny any good prayer. Each answer will of course come in GOD'S perfect time and way. But GOD *is* bound in their immutable perfection to grant every eternal request, born of eternal love. The eternal perspective enables faster higher understanding, allowing us to communicate intimately and responsibly with GOD, and know what prayers are amiss. Many gravitate to a religious hope that GOD loves everyone for eternity; and gives everyone everything they physically desire to make them eternally happy. But what measure of eternal love between them and GOD is found in that? That sounds like eternal individual division to me. The true model is a shared spiritual oneness of desire. Now I learned early on that my prior understanding of reality, found through a temporary perspective was all wrong. It had all been formed of worldly ways, teachings, and desires, and needed to be replaced. I did surrender my every worldly understanding and belief before GOD; well actually, there I strongly began my journey of doing so. This is a founding aspect of the eternal perspective, which in my moment of surrender, I knew nothing about. Did I stumble upon these aspects of the eternal perspective by luck? No. I sensed I was being led, and trusted in faith that the first foundational *stones* that support my perspective were right and true. And 47+ years on, GOD has built *my* wall of eternal perspective in clarity, strength, and understanding, from what seems to be endless perfect stones. If we love GOD above all, and ask for any eternal understanding, GOD will build a way to bring that perfect understanding to us; which so often require certain other understandings must first be taught and learned, for us to hear and understand what we seek. We must also realize, that without a true core desire for GOD, we cannot ask GOD for something good and eternal as a skeptical test of their existence; GOD is not mocked or tested. Now another question of vast importance is, what are the Scriptures? The Scriptures are a map, revealing layers of truth that to hear, depends on where and who we currently are spiritually. Starting our journey towards GOD in a true way, we see Scriptures are divided by HOLYSPIRIT into two parts. First, the Creation of reality and its fall, before CHRIST'S sacrifice; second, CHRIST'S sacrifice, and the coming end of it all. The physical life, death, and resurrection of JESUS, was the greatest eternal moment of love, to ever occur within this created reality. And by all accountability, it defines the final efforts of GOD to reach, teach, and show mankind what caused it, and what must come, after such a profound occurrence of sacrificial love. Scripture is our ancient historical chronicle of mankind's free-will responses to all of GOD'S attempts to call, teach, and re-endow souls with the eternal perspective. It is the account of how one third of all angels, and most all mankind, rejected this perspective in their free-will choices. Scripture is seasoned throughout with pearls of wisdom, and gems of insight into what GOD has done, and will do, and why. These are called the former, and as yet, unseen prophetic fulfillments; do not doubt these. Scripture chronicles the ordered events, and the scientific layers of the original creation process; and how that once perfect creation, fell away from GOD. It is the account of GOD'S many attempts to forgive; and GOD being repeatedly betrayed. Yet the greatest loving sacrifice to exist was still given. It is the account of the great pains and sufferings of GOD, the Angels, and Mankind, to find between them a family of free-willed love that will endure for eternity. A wonder of love that can only endure through GOD'S and all our faithful responsibility to protect and perpetuate the same.

Finding *and* preserving eternal love is the only objective of the eternal perspective. Eternal love can never be known, held, or forged through a temporary perspective. For all souls eternally under GOD, there can only be one mutually shared perspective, seeing GOD will not sustain countless complex realities, to satisfy all individual perspectives. Souls in their endless individual temporary perspectives want to be heard, understood, and accepted for who and what *they* have desired and chosen to be; they refuse to see how the only true way above all things, is to lovingly hear, understand, and accept who GOD is, and gifted and created us to be. We are not the creators of reality, or its sustaining truth, but we *can* hear, learn, understand, and accept GOD'S truth. Scripture at the very end is charged with the Great Prophecy of CHRIST. A Greatest Revelation, being a chronological summation of coming accountable destruction and judgment, leading up to that final eternal separation. These are the most powerful fearful writings in all reality. Why? Because these are not only going to occur in the exact order in which they are written; but the world will still not read, or believe, and the world will deny it as it all occurs. They will reap what they each sow in their ardent confused arrogance. The Judgments of GOD will flow over the lost in wave after wave, as their imaginations try to re-define for their selves what, and why it is happening. Most will blame corporate and political AI deception, pollution, biological research gone astray, and war. Some see cosmic forces going astray, or extraterrestrial sentient interference. But we will hear much ridicule about the true cause. Governments, the sciences, news outlets, celebrities, and false religions, around the world will make sport, and shame all who speak about the Judgment of GOD. And most souls will laugh at all who know GOD is judging all evil ways. The mysteries of Scripture can't be outgrown, as they are our foundation and gateway to higher spiritual wisdom and phenomena. The HOLYSPIRIT and Scripture together, are GOD'S channel of spiritual communication and revelation with us. There is nothing FATHER cannot open to us through CHRIST, HOLYSPIRIT, and Scripture. If the Bible is not continually opening up to us, we are failing to open to it. Open yourself to every step of growth in the spiritual metamorphous laid before us. Never find comfort in where you are, until you are with GOD; till then, always want more of, and to be closer to GOD. But never doubt, never say we cannot find, never let go. If we ever say we can't, we are letting go of the very power of GOD that says we can. Never be deceived about the reason that any soul lets go of

GOD; I tell you none let go unless they truly want to, in losing their strength of love to hold on. Lines, lines, lines, are consciously and spiritually drawn by everyone, everywhere, in everything, and in all moments of life within FATHER'S created reality. But who draws, and far more importantly, who has the right to draw the lines that pertain to life? And what are the eternal results, if everyone draws their own lines? Or if GOD draws these lines pertaining to life? We see hell and death, or eternal life and love. In all our life's desires and choices, we all have our current certain capacity to love. Life's evil choices and desires chip away at this ability within us; but how much can we be repaired. If our over-all ability and capacity to love has been severely damaged, we may not be able to love GOD in an adequate eternal way; so not being reprobate, yct too diminished to eternally and responsibly love GOD and life. But we can never know this, so never give up; strive, run, and climb towards GOD with all you are. If any ever do completely let go of GOD, they cannot see their self as the only reason; at least not for long. They will go more insane, and begin to believe their own made-up reasons as they fall and sink into the abyss of hell's confusion. The eternal perspective of reality is so amazing; it is our grounding and anchoring understandings of joy, love, peace, and family, held by one common belief. Some have it, and yet do not know what it is. A temporary perspective is a dwindling sense that turns more and more inward, becoming more and more separated from everyone and everything beyond their soul. Imagine being forever in the conscious isolation of an anguished lonely confused belief, thinking I don't know, it makes no sense, where am I. Endless fearful thought after thought, not bound to any eternal foundation, having no grasp on any shared sustained safe belief in reality. Being an unsecured soul, roiling forever in a state of chosen individual ever-changing insanity. What is insanity? Insanity is simply not knowing and being grounded to GOD'S true created reality. GOD tells us they are our only mooring hope for a sustained reality. GOD lovingly hopes for nothing less than to provide us the ways and place where we can all hang on to each other in one eternal understanding of love. A materialist view is full of paradoxes, as the complexities of all *basic* energy and mass have no eternal purpose to exist; and all their declared agents of change are destructive, not creative. Not one aspect of energy or mass is creative in itself; only force creates and destroys. And in their theories, they have no point of origin, or first generation for force; they just assume force has always been. This fact alone, shows their founding physical miracles they build upon, lack any true vision or understanding. Even the most *basic* forms of energy or mass sensed by mankind, have deep complexities, hidden just there beyond their comprehension. Complexities that show their theories and beliefs, see no true foundation. If fallen mankind cannot in a full and relative way, define the true first foundation of any one thing; they cannot begin to declare that entangled complex thing itself. Through GOD, we can see this fallen reality from its entangled complex beginning, all the way to its entangled end, without one unknown aspect or paradox. Reality's one true understanding resides in pure consonance for us. We have no flaws or weakness or missing requirements in our understanding of it all. We see only pure perfect strength, truth, reason, and purpose; we see only love and GOD. Open your hearts to see what is truly there; lovingly embrace all of this, and it will all lovingly embrace you. Look up and out beyond time and all other temporary illusionary dimensions and constraints they have set upon your soul, and see why you can never cease to exist as a conscious free-willed forceful entity. We can never completely lose our knowing awareness that we exist. But from a temporary devolving perspective, one can shrink till that is all they will ever know in sorrow and confusion. The creator of truth and reality, declares that from one eternal *shared* perspective, there is no end to what we can all *truly* know and share together. But how can a soul learn, what they refuse to learn? How can a soul know, what they refuse to know? How can a soul hear and see, what they refuse to hear and see? How can a soul want, what they don't want? This is GOD'S and every teacher's frustration and dilemma. Most souls, just do not want the fullness of GOD, or any true understanding; many want their desired version of GOD and truth. To see and truly know GOD, do we first see GOD'S physical personage, and then their spiritual attributes? No. We first see their one truth and eternal ways of life and love, springing forth as the spiritual aspects of their desires. All must understand that since there is one ruling truth to reality, that long preceded *us*, we can only be led to it, or wander from it forever. How impossible would it be for fallen mankind to find GOD'S one truth on their own? Where would everyone start? How could anyone know it is the truth if GOD does not share it with us? Everyone would go in different directions, looking in different places, finding different truths; just as they now do. The true understanding, and the one eternal loving way of life, are as much a part of GOD, as are the eternal forms and visage they indwell. They have been one together, as GOD is one and unchanging in all spiritual loving aspects since this reality's true beginning. Remember that any soul, including ours, can fall from GOD by the force of *one* strong evil desire hiding within us. No soul can rise up to be with GOD, unless there are two forces acting upon us: one being, our found and nurtured forceful loving desire *for* GOD and their Family pushing us upwards; and two, the forceful loving desire *from* GOD and Family pulling us upward. Most souls can't comprehend just how far this world and all that is in it have fallen from GOD; and it is so, so far. We see how everything fallen mankind tries to understand and improve life, but they only make it worse and destroy it; and the same for the ways of the world around them. Many say the U.S. has no culture; they say this because they can't see independent self-reliant freedom and bravery as a culture. A socialist/communist mind, drains their home countries of wealth, ambition, and spirit, and flee here to do the same; they are parasitic heterotrophs. All understanding of compassion is unbalanced in them. They hold only disdain for those who achieve and provide, and all their compassion is for those who don't. Wealth rises from unbridled ingenuity and ambition, and not from Government control; our culture is freedom. We know we cannot ask or expect anything of, or from GOD, if we are not giving to GOD all the things they need and expect from us; for this is love. Love must balance, or one side is always hurt; and there shall be no *new* pain in Heaven. Now for mass consumption, we argue all our points in all our books from a certain simple level; but please know that we find GOD'S model always holds true, for as long and deep as any care to argue against it. From our eternal perspective, we have never found GOD'S model to be questioned or diminished; for us, GOD'S model of reality is an immutable brute-force argument, once it is properly and fully seen and understood. Please hear me, if you are to overcome all divisions of evil within and around you, and become one with GOD, you must grasp the truth, that from our conception, we began to more and more sense and interact with the physical information of GOD'S Creation. But from our first moment, most souls began to burrow into, and bury themselves in that physical information, and the temporary ways and loves of this fallen world. Who can fully grasp and know their situation or position, and their state of contradiction, we all must escape from? Who can know the responsible faithful love required? We must run and not walk, we must strive and fight, and not be still or surrender. I tell you all I can, all I have seen, all I know, I share all my love and fear; yet we each have our own part, our race, our fight, our desires, and our loves and fears. Life is a precious eternal gift to be treasured, nourished, and protected. May GOD'S truth be your joy, your true love, and your true mandate; and if so, your burden will be as light as a feather, as CHRIST'S love will carry you over. Open your eyes, your heart, and your soul wide; wake up and sing praise to GOD above! Now if you were pulled back and forth in reading this; I hope you pray, fast, and read the Bible and our books again. GOD righteously prepares a place for the life perpetuating forces of truth. They also prepare a place for the life dividing forces of darkness. One being eternally separated from his wrath, by his great love. One being eternally separated from GOD'S love by his wrath. Great joy or sorrow, we will all too soon be in one of these two places. Let GOD have mercy upon you and me and many, knowing he will end this soon. Amen. If you should desire to share GOD'S love or helpful *positive* critique, or ask us a true meaningful question. Or if you can't afford and need a reduced-price copy of this, or any of our other three books listed on page two, for yourself or someone else who cannot afford it, then we want to try to help. So please, for any and all of these reasons above, email Michael @ livetoknowtheGODofprophecy@gmail.com